U.S. Army Survival Series

The Complete U.S. Army Survival Guide to

FORAGING

Skills, Tactics, and Techniques

U.S. Army Survival Series

The Complete U.S. Army Survival Guide to

FORAGING

Skills, Tactics, and Techniques

Edited by

Jay McCullough

Skyhorse Publishing

Skyhorse Publishing books may be purchased in bulk at special discounts for sales promotion, corporate gifts, fund-raising, or educational purposes. Special editions can also be created to specifications. For details, contact the Special Sales Department, Skyhorse Publishing, 307 West 36th Street, 11th Floor, New York, NY 10018 or info@skyhorsepublishing.com.

Skyhorse® and Skyhorse Publishing® are registered trademarks of Skyhorse Publishing, Inc.®, a Delaware corporation.

Visit our website at www.skyhorsepublishing.com.

10 9 8 7 6 5 4

Library of Congress Cataloging-in-Publication Data is available on file.

Cover design by Tom Lau

Print ISBN: 978-1-5107-0743-6
Ebook ISBN: 978-1-5107-0748-1

Printed in the United States of America

Table of Contents

Introduction

When it comes to survival foraging, there is no substitute for previous research and practical experience. And when it comes to previous research and practical experience for advice on what is edible and what is not, there is no substitute for the accumulated wisdom of people who have been foraging for decades, especially indigenous peoples.

A particularly good example of this is the practice of identifying edible and poisonous mushrooms. A cursory search on matters mycological will yield twenty warnings as serious as a heart attack, all advising you to find a certified local mushroom guru for every singular piece of genuinely useful information. I'm inclined to agree on this point, as a small number of mushroom admirers are certain to win the Darwin Award every year.

In this book of materials culled from the U.S. Army, you will find an excellent basis for the study of survival foraging, touching on a variety of edibles and nasties from around the globe. If you anticipate being left on a spit of God's own desolation and think you'll have to eat your way out, I would recommend that you commit much of it to memory for starters; I doubt that the *Complete U.S. Army Survival Guide to Foraging Skills, Tactics, and Techniques* will drop out of the sky in your hour of need.

The next step is to make a detailed, first-hand study of the flora and fauna of your region of interest. Since everything is seasonal, this study would ideally be over the course of at least a year. I think you will find that it will be possible to live off the fat of the land in some cases, and that you would likely starve in other cases. But it may be sufficient to use foraging to stave off hunger for as long as it takes to be rescued or reach a food source.

In any case, when it comes to survival foraging, there is no substitute for previous research and practical experience.

—Jay McCullough
February 2016
North Haven, Connecticut

CHAPTER 1

WATER PROCUREMENT

Water is one of your most urgent needs in a survival situation. You can't live long without it, especially in hot areas where you lose water rapidly through perspiration. Even in cold areas, you need a minimum of 2 liters of water each day to maintain efficiency.

More than three-fourths of your body is composed of fluids. Your body loses fluid as a result of heat, cold, stress, and exertion. To function effectively, you must replace the fluid your body loses. So, one of your first goals is to obtain an adequate supply of water.

WATER SOURCES

Almost any environment has water present to some degree. Table 1-1 lists possible sources of water in various environments. It also provides information on how to make the water potable.

Note: If you do not have a canteen, a cup, a can, or other type of container, improvise one from plastic or water-resistant cloth. Shape the plastic or cloth into a bowl by pleating it. Use pins or other suitable items—even your hands—to hold the pleats.

If you do not have a reliable source to replenish your water supply, stay alert for ways in which your environment can help you.

> ⚠ CAUTION
> Do not substitute the fluids listed in Table 1-2 for water.

Heavy dew can provide water. Tie rags or tufts of fine grass around your ankles and walk through dew-covered grass before sunrise. As the rags or grass tufts absorb the dew, wring the water into a container. Repeat the process until you have a supply of water or until the dew is gone. Australian natives sometimes mop up as much as a liter an hour this way.

Table 1-1: Water sources in different environments

Environment	Source of Water	Means of Obtaining and/or Making Potable	Remarks
Frigid areas	Snow and ice	Melt and purify.	Do not eat without melting! Eating snow and ice can reduce body temperature and will lead to more dehydration.
			Snow and ice are no purer than the water from which they come.
			Sea ice that is gray in color or opaque is salty. Do not use it without desalting it. Sea ice that is crystalline with a bluish cast has little salt in it.
At sea	Sea	Use desalter kit.	Do not drink seawater without desalting.
	Rain	Catch rain in tarps or in other water-holding material or containers.	If tarp or water-holding material has become encrusted with salt, wash it in the sea before using (very little salt will remain on it).
	Sea ice		See remarks above for frigid areas.

Table 1-1: *(Continued)*

Environment	Source of Water	Means of Obtaining and/or Making Potable	Remarks
Beach	Ground	Dig hole deep enough to allow water to seep in; obtain rocks, build fire, and heat rocks; drop hot rocks in water; hold cloth over hole to absorb steam; wring water from cloth.	Alternate method if a container or bark pot is available: Fill container or pot with seawater; build fire and boil water to produce steam; hold cloth over container to absorb steam; wring water from cloth.
Desert	Ground • in valleys and low areas • at foot of concave banks of dry river beads • at foot of cliffs or rock outcrops • at first depression behind first sand dune of dry desert lakes • wherever you find damp surface sand • wherever you find green vegetation	Dig holes deep enough to allow water to seep in.	In a sand dune belt, any available water will be found beneath the original valley floor at the edge of dunes.
	Cacti	Cut off the top of a barrel cactus and mash or squeeze the pulp. CAUTION: Do not eat pulp. Place pulp in mouth, suck out juice, and discard pulp.	Without a machete, cutting into a cactus is difficult and takes time since you must get past the long, strong spines and cut through the tough rind.

Table 1-1: *(Continued)*

Environment	Source of Water	Means of Obtaining and/or Making Potable	Remarks
Desert (continued)	Depressions or holes in rocks	Dig hole deep enough to allow water to seep in; obtain rocks, build	Periodic rainfall may collect in pools, seep into fissures, or collect in holes in rocks.
	Fissures in rock	Insert flexible tubing and siphon water. If fissure is large enough, you can lower a container into it.	
	Porous rock	Insert flexible tubing and siphon water.	
	Condensation on metal	Use cloth to absorb water, then wring water from cloth.	Extreme temperature variations between night and day may cause condensation on metal surfaces.
			Following are signs to watch for in the desert to help you find water:
			• All trails lead to water. You should follow in the direction in which the trails converge. Signs of camps, campfire ashes, animal droppings, and trampled terrain may mark trails.
			• Flocks of birds will circle over water holes. Some birds fly to water holes at dawn and sunset. Their flight at these times is generally fast and close to the ground. Bird tracks or chirping sounds in the evening or early morning sometimes indicate that water is nearby.

Table 1-2: The effects of substitute fluids

Fluid	Remarks
Alcoholic beverages	Dehydrate the body and cloud judgment.
Urine	Contains harmful body wastes. Is about 2 percent salt.
Blood	Is salty and considered a food; therefore, requires additional body fluids to digest. May transmit disease.
Seawater	Is about 4 percent salt. It takes about 2 liters of body fluids to rid the body of waste from 1 liter of seawater. Therefore, by drinking seawater you deplete your body's water supply, which can cause death.

Bees or ants going into a hole in a tree may point to a water-filled hole. Siphon the water with plastic tubing or scoop it up with an improvised dipper. You can also stuff cloth in the hole to absorb the water and then wring it from the cloth.

Water sometimes gathers in tree crotches or rock crevices. Use the above procedures to get the water. In arid areas, bird droppings around a crack in the rocks may indicate water in or near the crack.

Green bamboo thickets are an excellent source of fresh water. Water from green bamboo is clear and odorless. To get the water, bend a green bamboo stalk, tie it down, and cut off the top (Figure 1-1). The water will drip freely during the night. Old, cracked bamboo may contain water.

⚠ CAUTION
Purify the water before drinking it.

Wherever you find banana or plantain trees, you can get water. Cut down the tree, leaving about a 30-centimeter stump, and scoop out the center of the stump so that the hollow is bowl-shaped. Water from the roots will immediately start to fill the hollow. The first three fillings of water will be bitter, but succeeding fillings will be palatable. The stump (Figure 1-2) will supply water for up to four days. Be sure to cover it to keep out insects.

Some tropical vines can give you water. Cut a notch in the vine as high as you can reach, then cut the vine off close to the ground. Catch the dropping liquid in a container or in your mouth (Figure 1-3).

⚠ CAUTION
Do not drink the liquid if it is sticky, milky, or bitter tasting.

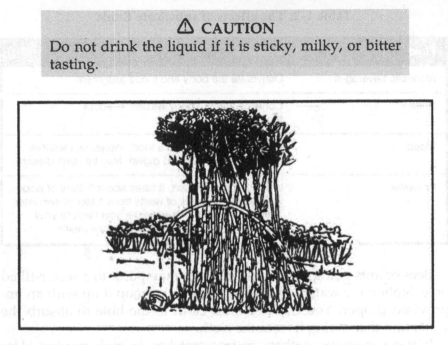

Figure 1-1: Water from green bamboo

The milk from green (unripe) coconuts is a good thirst quencher. However, the milk from mature coconuts contains an oil that acts as a laxative. Drink in moderation only.

In the American tropics you may find large trees whose branches support air plants. These air plants may hold a considerable amount of rainwater in their overlapping, thickly growing leaves. Strain the water through a cloth to remove insects and debris.

You can get water from plants with moist pulpy centers. Cut off a section of the plant and squeeze or smash the pulp so that the moisture runs out. Catch the liquid in a container.

Plant roots may provide water. Dig or pry the roots out of the ground, cut them into short pieces, and smash the pulp so that the moisture runs out. Catch the liquid in a container.

Fleshy leaves, stems, or stalks, such as bamboo, contain water. Cut or notch the stalks at the base of a joint to drain out the liquid.

The following trees can also provide water:

- Palms. Palms, such as the buri, coconut, sugar, rattan, and nips, contain liquid. Bruise a lower frond and pull it down so the tree will "bleed" at the injury.

CUT HERE

CUT OUT BOWL

Water will fill bowl from roots.

Figure 1-2: Water from plantain or banana tree stump

- Traveler's tree. Found in Madagascar, this tree has a cuplike sheath at the base of its leaves in which water collects.
- Umbrella tree. The leaf bases and roots of this tree of western tropical Africa can provide water.
- Baobab tree. This tree of the sandy plains of northern Australia and Umbrella tree. The leaf bases and roots of this tree of western tropical Africa can provide water.

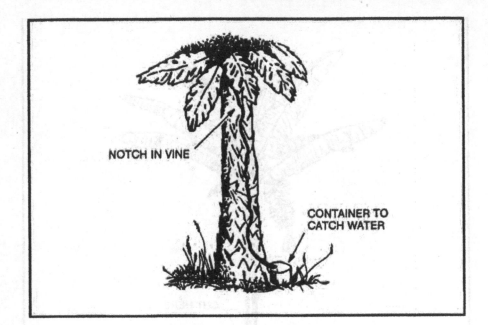

NOTCH IN VINE

CONTAINER TO
CATCH WATER

Figure 1-3: Water from a vine

> ⚠ CAUTION
> Do not keep the sap from plants longer than 24 hours.
> It begins fermenting, becoming dangerous as a water
> source.

STILL CONSTRUCTION

You can use stills in various areas of the world. They draw moisture
from the ground and from plant material. You need certain materials
to build a still, and you need time to let it collect the water. It takes
about 24 hours to get 0.5 to 1 liter of water.

Aboveground Still
To make the aboveground still, you need a sunny slope on which to
place the still, a clear plastic bag, green leafy vegetation, and a small
rock (Figure 1-4).

To make the still—

- Fill the bag with air by turning the opening into the breeze or by "scooping" air into the bag.
- Fill the plastic bag half to three-fourths full of green leafy vegetation. Be sure to remove all hard sticks or sharp spines that might puncture the bag.

⚠ CAUTION

Do not use poisonous vegetation. It will provide poisonous liquid.

- Place a small rock or similar item in the bag.
- Close the bag and tie the mouth securely as close to the end of the bag as possible to keep the maximum amount of air space. If you have a piece of tubing, a small straw, or a hollow reed, insert one end in the mouth of the bag before you

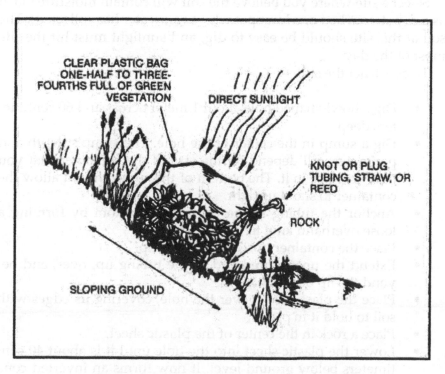

CLEAR PLASTIC BAG
ONE-HALF TO THREE-
FOURTHS FULL OF GREEN
VEGETATION

DIRECT SUNLIGHT

KNOT OR PLUG
TUBING, STRAW, OR
REED

ROCK

SLOPING GROUND

Figure 1-4: Aboveground solar water still

tie it securely. Then tie off or plug the tubing so that air will not escape. This tubing will allow you to drain out condensed water without untying the bag.

- Place the bag, mouth downhill, on a slope in full sunlight. Position the mouth of the bag slightly higher than the low point in the bag.
- Settle the bag in place so that the rock works itself into the low point in the bag.

To get the condensed water from the still, loosen the tie around the bag's mouth and tip the bag so that the water collected around the rock will drain out. Then retie the mouth securely and reposition the still to allow further condensation.

Change the vegetation in the bag after extracting most of the water from it. This will ensure maximum output of water.

Belowground Still

To make a belowground still, you need a digging tool, a container, a clear plastic sheet, a drinking tube, and a rock (Figure 1-5).

Select a site where you believe the soil will contain moisture (such as a dry stream bed or a low spot where rainwater has collected). The soil at this site should be easy to dig, and sunlight must hit the site most of the day.

To construct the still—

- Dig a bowl-shaped hole about 1 meter across and 60 centimeters deep.
- Dig a sump in the center of the hole. The sump's depth and perimeter will depend on the size of the container that you have to place in it. The bottom of the sump should allow the container to stand upright.
- Anchor the tubing to the container's bottom by forming a loose overhand knot in the tubing.
- Place the container upright in the sump.
- Extend the unanchored end of the tubing up, over, and beyond the lip of the hole.
- Place the plastic sheet over the hole, covering its edges with soil to hold it in place.
- Place a rock in the center of the plastic sheet.
- Lower the plastic sheet into the hole until it is about 40 centimeters below ground level. It now forms an inverted cone

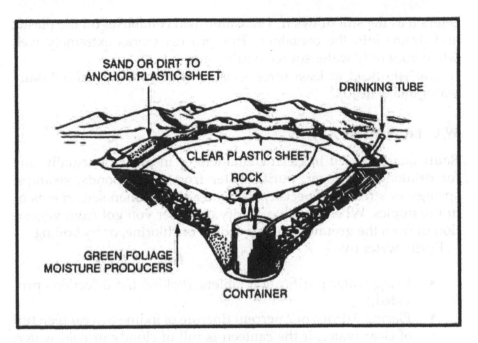

SAND OR DIRT TO
ANCHOR PLASTIC SHEET

DRINKING TUBE

CLEAR PLASTIC SHEET

ROCK

GREEN FOLIAGE
MOISTURE PRODUCERS

CONTAINER

Figure 1-5: Belowground still

with the rock at its apex. Make sure that the cone's apex is
directly over your container. Also make sure the plastic cone
does not touch the sides of the hole because the earth will absorb the condensed water.
  • Put more soil on the edges of the plastic to hold it securely in
    place and to prevent the loss of moisture.
  • Plug the tube when not in use so that the moisture will not
    evaporate.

You can drink water without disturbing the still by using the tube as
a straw.
    You may want to use plants in the hole as a moisture source. If so,
dig out additional soil from the sides of the hole to form a slope on
which to place the plants. Then proceed as above.
    If polluted water is your only moisture source, dig a small trough
outside the hole about 25 centimeters from the still's lip (Figure 1-6).
Dig the trough about 25 centimeters deep and 8 centimeters wide.
Pour the polluted water in the trough. Be sure you do not spill any
polluted water around the rim of the hole where the plastic sheet
touches the soil. The trough holds the polluted water and the soil

filters it as the still draws it. The water then condenses on the plastic and drains into the container. This process works extremely well when your only water source is salt water.

You will need at least three stills to meet your individual daily water intake needs.

WATER PURIFICATION

Rainwater collected in clean containers or in plants is usually safe for drinking. However, purify water from lakes, ponds, swamps, springs, or streams, especially the water near human settlements or in the tropics. When possible, purify all water you got from vegetation or from the ground by using iodine or chlorine, or by boiling.

Purify water by—

- Using water purification tablets. (Follow the directions provided.)
- Placing 5 drops of 2 percent tincture of iodine in a canteen full of clear water. If the canteen is full of cloudy or cold water, use 10drops. (Let the canteen of water stand for 30 minutes before drinking.)
- Boiling water for 1 minute at sea level, adding 1 minute for each additional 300 meters above sea level, or boil for 10 minutes no matter where you are.

By drinking nonpotable water you may contract diseases or swallow organisms that can harm you. Examples of such diseases or organisms are—

- Dysentery. Severe, prolonged diarrhea with bloody stools, fever, and weakness.
- Cholera and typhoid. You may be susceptible to these diseases regardless of inoculations.
- Flukes. Stagnant, polluted water—especially in tropical areas—often contains blood flukes. If you swallow flukes, they will bore into the bloodstream, live as parasites, and cause disease.
- Leeches. If you swallow a leech, it can hook onto the throat passage or inside the nose. It will suck blood, create a wound, and move to another area. Each bleeding wound may become infected.

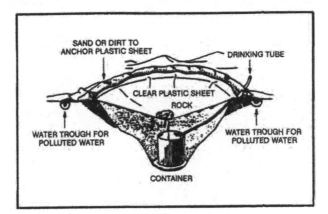

Figure 1-6: Belowground still to get potable water from polluted water

WATER FILTRATION DEVICES

If the water you find is also muddy, stagnant, and foul smelling, you can clear the water—

- By placing it in a container and letting it stand for 12 hours.
- By pouring it through a filtering system.

Note: These procedures only clear the water and make it more palatable. You will have to purify it.

To make a filtering system, place several centimeters or layers of filtering material such as sand, crushed rock, charcoal, or cloth in bamboo, a hollow log, or an article of clothing (Figure 1-7).

Remove the odor from water by adding charcoal from your fire. Let the water stand for 45 minutes before drinking it.

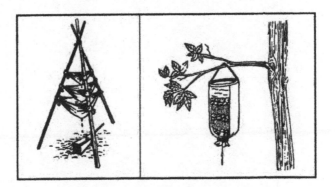

Figure 1-7: Water filtering systems

CHAPTER 2

FOOD PROCUREMENT

After water, man's most urgent requirement is food. In contemplating virtually any hypothetical survival situation, the mind immediately turns to thoughts of food. Unless the situation occurs in an arid environment, even water, which is more important to maintaining body functions, will almost always follow food in our initial thoughts. The survivor must remember that the three essentials of survival—water, food, and shelter–are prioritized according to the estimate of the actual situation. This estimate must not only be timely but accurate as well. Some situations may well dictate that shelter precede both food and water.

ANIMALS FOR FOOD

Unless you have the chance to take large game, concentrate your efforts on the smaller animals, due to their abundance. The smaller animal species are also easier to prepare. You must not know all the animal species that are suitable as food. Relatively few are poisonous, and they make a smaller list to remember. What is important is to learn the habits and behavioral patterns of classes of animals. For example, animals that are excellent choices for trapping, those that inhabit a particular range and occupy a den or nest, those that have somewhat fixed feeding areas, and those that have trails leading from one area to another. Larger, herding animals, such as elk or caribou, roam vast areas and are somewhat more difficult to trap. Also, you must understand the food choices of a particular species.

You can, with relatively few exceptions, eat anything that crawls, swims, walks, or flies. The first obstacle is overcoming your natural aversion to a particular food source. Historically, people in starvation situations have resorted to eating everything imaginable for nourishment. A person who ignores an otherwise healthy food source due to a personal bias, or because he feels it is unappetizing, is risking his own survival. Although it may prove difficult at first, a survivor must eat what is available to maintain his health.

Insects

The most abundant life-form on earth, insects are easily caught. Insects provide 65 to 80 percent protein compared to 20 percent for beef. This fact makes insects an important, if not overly appetizing, food source. Insects to avoid include all adults that sting or bite, hairy or brightly colored insects, and caterpillars and insects that have a pungent odor. Also avoid spiders and common disease carriers such as ticks, flies, and mosquitoes. See Chapter 3 for more information about dangerous insects and arachnids.

Rotting logs lying on the ground are excellent places to look for a variety of insects including ants, termites, beetles, and grubs, which are beetle larvae. Do not overlook insect nests on or in the ground. Grassy areas, such as fields, are good areas to search because the insects are easily seen. Stones, boards, or other materials lying on the ground provide the insects with good nesting sites. Check these sites. Insect larvae are also edible. Insects such as beetles and grasshoppers that have a hard outer shell will have parasites. Cook them before eating. Remove any wings and barbed legs also. You can eat most insects raw. The taste varies from one species to another. Wood grubs are bland, while some species of ants store honey in their bodies, giving them a sweet taste.

You can grind a collection of insects into a paste. You can mix them with edible vegetation. You can cook them to improve their taste.

Worms

Worms (Annelidea) are an excellent protein source. Dig for them in damp humus soil or watch for them on the ground after a rain. After capturing them, drop them into clean, potable water for a few minutes. The worms will naturally purge or wash themselves out, after which you can eat them raw.

Leeches

Leeches are blood-sucking creatures with a wormlike appearance. You find them in the tropics and in temperate zones. You will certainly encounter them when swimming in infested waters or making expedient water crossings. You can find them when passing through swampy, tropical vegetation and bogs. You can also find them while cleaning food animals, such as turtles, found in fresh water. Leeches can crawl into small openings; therefore, avoid camping in their habitats when possible. Keep your trousers tucked in your boots. Check yourself frequently for leeches. Swallowed or eaten, leeches can be a great hazard. It is therefore essential to treat water from questionable sources by boiling or using chemical water treatments. Survivors have developed severe infections from wounds inside the throat or nose when sores from swallowed leeches became infected.

Crustaceans

Freshwater shrimp range in size from 0.25 centimeter up to 2.5 centimeters. They can form rather large colonies in mats of floating algae or in mud bottoms of ponds and lakes.

Crayfish are akin to marine lobsters and crabs. You can distinguish them by their hard exoskeleton and five pairs of legs, the front pair having oversized pincers. Crayfish are active at night, but you can locate them in the daytime by looking under and around stones in streams. You can also find them by looking in the soft mud near the chimney-like breathing holes of their nests. You can catch crayfish by tying bits of offal or internal organs to a string. When the cray fish grabs the bait, pull it to shore before it has a chance to release the bait.

You find saltwater lobsters, crabs, and shrimp from the surf's edge out to water 10 meters deep. Shrimp may come to a light at night where you can scoop them up with a net. You can catch lobsters and crabs with a baited trap or a baited hook. Crabs will come to bait placed at the edge of the surf, where you can trap or net them. Lobsters and crabs are nocturnal and caught best at night.

Mollusks

This class includes octopuses and freshwater and saltwater shell fish such as snails, clams, mussels, bivalves, barnacles, periwinkles, chitons, and sea urchins (Figure 2-1). You find bivalves similar to our freshwater mussel and terrestrial and aquatic snails worldwide under all water conditions.

River snails or freshwater periwinkles are plentiful in rivers, streams, and lakes of northern coniferous forests. These snails may be pencil point or globular in shape.

In fresh water, look for mollusks in the shallows, especially in water with a sandy or muddy bottom. Look for the narrow trails they leave in the mud or for the dark elliptical slit of their open valves.

Near the sea, look in the tidal pools and the wet sand. Rocks along beaches or extending as reefs into deeper water often bear clinging shellfish. Snails and limpets cling to rocks and seaweed from the low

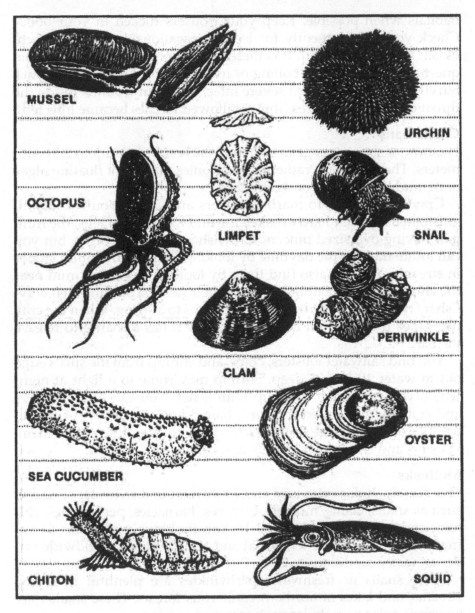

Figure 2-1: Edible Mollusks

water mark upward. Large snails, called chitons, adhere tightly to rocks above the surf line.

Mussels usually form dense colonies in rock pools, on logs, or at the base of boulders.

⚠ CAUTION

Mussels may be poisonous in tropical zones during the summer!

Steam, boil, or bake mollusks in the shell. They make excellent stews in combination with greens and tubers.

⚠ CAUTION

Do not eat shellfish that are not covered by water at high tide!

See chapter 5 for more information about dangerous shellfish.

Fish

Fish represent a good source of protein and fat. They offer some distinct advantages to the survivor or evader. They are usually more abundant than mammal wildlife, and the ways to get them are silent. To be successful at catching fish, you must know their habits. For instance, fish tend to feed heavily before a storm. Fish are not likely to feed after a storm when the water is muddy and swollen. Light often attracts fish at night. When there is a heavy current, fish will rest in places where there is an eddy, such as near rocks. Fish will also gather where there are deep pools, under overhanging brush, and in and around submerged foliage, logs, or other objects that offer them shelter.

There are no poisonous freshwater fish. However, the catfish species has sharp, needlelike protrusions on its dorsal fins and barbels. These can inflict painful puncture wounds that quickly become infected.

Cook all freshwater fish to kill parasites. Also cook saltwater fish caught within a reef or within the influence of a freshwater source as a precaution. Any marine life obtained farther out in the sea will not contain parasites because of the saltwater environment. You can eat these raw.

Many fish living in reefs near shore, or in lagoons and estuaries, are poisonous to eat, though some are only seasonally dangerous. The majority are tropical fish; however, be wary of eating any unidentifiable fish wherever you are. Some predatory fish, such as barracuda and snapper, may become toxic if the fish they feed on in shallow waters are poisonous. The most poisonous types appear to have parrotlike beaks and hard shell-like skins with spines and often can inflate their bodies like balloons. However, at certain times of the year, indigenous populations consider the puffer a delicacy. See chapter 5: Dangerous Fish, Mollusks, and Freshwater Animals for more information.

Amphibians

Frogs and salamanders are easily found around bodies of fresh water. Frogs seldom move from the safety of the water's edge. At the first sign of danger, they plunge into the water and bury themselves in the mud and debris. There are few poisonous species of frogs. Avoid any brightly colored frog or one that has a distinct "X" mark on its back. Do not confuse toads with frogs. You normally find toads in drier environments. Several species of toads secrete a poisonous substance through their skin as a defense against attack. Therefore, to avoid poisoning, do not handle or eat toads.

Salamanders are nocturnal. The best time to catch them is at night using a light. They can range in size from a few centimeters to well over 60 centimeters in length. Look in water around rocks and mud banks for salamanders.

Reptiles

Reptiles are a good protein source and relatively easy to catch. You should cook them, but in an emergency, you can eat them raw. Their raw flesh may transmit parasites, but because reptiles are cold-blooded, they do not carry the blood diseases of the warm-blooded animals. Another toxic meat is the flesh of the hawksbill turtle. You recognize them by their down-turned bill and yellow polka dots on their neck and front flippers. They weigh more than 275 kilograms and are unlikely to be captured.

The box turtle is a commonly encountered turtle that you should not eat. It feeds on poisonous mushrooms and may build up a highly toxic poison in its flesh. Cooking does not destroy this toxin. Avoid the hawksbill turtle, found in the Atlantic Ocean, because of its poisonous thorax gland. Poisonous snakes, alligators, crocodiles, and large sea turtles present obvious hazards to the survivor. See chapter 4 for more information about dangerous reptiles.

Birds

All species of birds are edible, although the flavor will vary considerably. You may skin fish-eating birds to improve their taste. As with any wild animal, you must understand birds' common habits to have

Table 2-1: Bird nesting places

Types of Birds	Frequent Nesting Places	Nesting Periods
Inland birds	Trees, woods, or fields	Spring and early summer in temperate and arctic regions; year round in the tropics
Cranes and herons	Mangrove swamps or high trees near water	Spring and early summer
Some species of owls	High trees	Late December through March
Ducks, geese, and swans	Tundra areas near ponds, rivers, or lakes	Spring and early summer in arctic regions
Some sea birds	Sandbars or low sand islands	Spring and early summer in temperate and arctic regions
Gulls, auks, murres, and cormorants	Steep rocky coasts	Spring and early summer in temperate and arctic regions

a realistic chance of capturing them. You can take pigeons, as well as some other species, from their roost at night by hand. During the nesting season, some species will not leave the nest even when approached. Knowing where and when the birds nest makes catching them easier (Table 2-1). Birds tend to have regular flyways going from the roost to a feeding area, to water, and so forth. Careful observation should reveal where these flyways are and indicate good areas for catching birds in nets stretched across the flyways (Figure 2-2). Roosting sites and waterholes are some of the most promising areas for trapping or snaring.

Nesting birds present another food source—eggs. Remove all but two or three eggs from the clutch, marking the ones at you leave. The bird will continue to lay more eggs to fill the clutch. Continue removing the fresh eggs, leaving the ones you marked.

Mammals
Mammals are excellent protein sources and, for Americans, the most tasty food source. There are some drawbacks to obtaining mammals. Ina hostile environment, the enemy may detect any traps or snares

placed on land. The amount of injury an animal can inflict is in direct proportion to its size. All mammals have teeth and nearly all will bite in self-defense. Even a squirrel can inflict a serious wound and any bite presents a serious risk of infection. Also, a mother can be extremely aggressive in defense of her young. Any animal with no route of escape will fight when cornered.

All mammals are edible; however, the polar bear and bearded seal have toxic levels of vitamin A in their livers. The platypus, native to Australia and Tasmania, is an egg-laying, semiaquatic mammal that has poisonous glands. Scavenging mammals, such as the opossum, may carry diseases. Common sense tells the survivor to avoid encounters with lions, bears, and other large or dangerous animals. You should also avoid large grazing animals with horns, hooves, and great weight. Your actions may prevent unexpected meetings. Move carefully through their environment. Do not attract large predators by leaving food lying around your camp. Carefully survey the scene before entering water or forests.

Bats

Despite the legends, bats (Desmodus species) are a relatively small hazard to the survivor. There are many bat varieties worldwide, but you find the true vampire bats only in Central and South America. They are small, agile fliers that land on their sleeping victims, mostly cows and horses, to lap a blood meal after biting their victim. Their saliva contains an anticoagulant that keeps the blood slowly flowing while they feed. Only a small percentage of these bats actually carry rabies; however, avoid any sick or injured bat. They can carry other diseases and infections and will bite readily when handled. Taking shelter in a cave occupied by bats, however, presents the much greater hazard of inhaling powdered bat dung, or guano. Bat dung carries many organisms that can cause diseases. Eating thoroughly cooked flying foxes or other bats presents no danger from rabies and other diseases, but again, the emphasis is on thorough cooking.

TRAPS AND SNARES

For an unarmed survivor or evader, or when the sound of a rifle shot could be a problem, trapping or snaring wild game is a good alternative. Several well-placed traps have the potential to catch much more game than a man with a rifle is likely to shoot. To be effective with any type of trap or snare, you must—

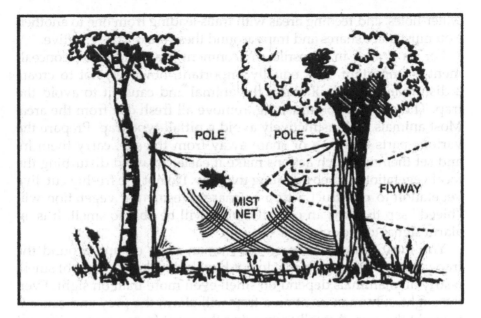

POLE ←   POLE →

FLYWAY

MIST
NET

Figure 2-2: Catching birds in a net

- Be familiar with the species of animal you intend to catch.
- Be capable of constructing a proper trap.
- Not alarm the prey by leaving signs of your presence.

There are no catchall traps you can set for all animals. You must determine what species are in a given area and set your traps specifically with those animals in mind. Look for the following

- Runs and trails.
- Tracks.
- Droppings.
- Chewed or rubbed vegetation.
- Nesting or roosting sites.
- Feeding and watering areas.

Position your traps and snares where there is proof that animals pass through. You must determine if it is a "run" or a "trail." A trail will show signs of use by several species and will be rather distinct. A run is usually smaller and less distinct and will only contain signs of one species. You may construct a perfect snare, but it will not catch anything if haphazardly placed in the woods. Animals have bedding areas, water-holes, and feeding areas with trails leading from one to another. You must place snares and traps around these areas to be effective.

For an evader in a hostile environment, trap and snare conceal-ment is important. It is equally important, however, not to create a disturbance that will alarm the animal and cause it to avoid the trap. Therefore, if you must dig, remove all fresh dirt from the area. Most animals will instinctively avoid a pitfall-type trap. Prepare the various parts of a trap or snare away from the site, carry them in, and set them up. Such actions make it easier to avoid disturbing the local vegetation, thereby alerting the prey. Do not use freshly cut, live vegetation to construct a trap or snare. Freshly cut vegetation will "bleed" sap that has an odor the prey will be able to smell. It is an alarm signal to the animal.

You must remove or mask the human scent on and around the trap you set. Although birds do not have a developed sense of smell, nearly all mammals depend on smell even more than on sight. Even the slightest human scent on a trap will alarm the prey and cause it to avoid the area. Actually removing the scent from a trap is difficult but masking it is relatively easy. Use the fluid from the gall and urine bladders of previous kills. Do not use human urine. Mud, particular-ly from an area with plenty of rotting vegetation, is also good. Use it to coat your hands when handling the trap and to coat the trap when setting it. In nearly all parts of the world, animals know the smell of burned vegetation and smoke. It is only when a fire is actually burn-ing that they become alarmed. Therefore, smoking the trap parts is an effective means to mask your scent. If one of the above techniques is not practical, and if time permits, allow a trap to weather for a few days and then set it. Do not handle a trap while it is weathering. When you position the trap, camouflage it as naturally as possible to prevent detection by the enemy and to avoid alarming the prey.

Traps or snares placed on a trail or run should use channelization. To build a channel, construct a funnel-shaped barrier extending from the sides of the trail toward the trap, with the narrowest part nearest the trap. Channelization should be inconspicuous to avoid alerting the prey. As the animal gets to the trap, it cannot turn left or right and continues into the trap. Few wild animals will back up, preferring to face the direction of travel. Channelization does not have to be an impassable barrier. You only have to make it inconvenient for the animal to go over or through the barrier. For best effect, the channel-ization should reduce the trail's width to just slightly wider than the targeted animal's body. Maintain this constriction at least as far back from the trap as the animal's body length, then begin the widening toward the mouth of the funnel.

Use of Bait

Baiting a trap or snare increases your chances of catching an animal. When catching fish, you must bait nearly all the devices. Success with an unbaited trap depends on its placement in a good location. A baited trap can actually draw animals to it. The bait should be something the animal knows. This bait, however, should not be so readily available in the immediate area that the animal can get it close by. For example, baiting a trap with corn in the middle of a corn field would not be likely to work. Likewise, if corn is not grown in the region, a corn-baited trap may arouse an animal's curiosity and keep it alerted while it ponders the strange food. Under such circumstances it may not go for the bait. One bait that works well on small mammals is the peanut butter from a meal, ready-to-eat (MRE) ration. Salt is also a good bait. When using such baits, scatter bits of it around the trap to give the prey a chance to sample it and develop a craving for it. The animal will then overcome some of its caution before it gets to the trap.

If you set and bait a trap for one species but another species takes the bait without being caught, try to determine what the animal was. Then set a proper trap for that animal, using the same bait.

Note: Once you have successfully trapped an animal, you will not only gain confidence in your ability, you also will have resupplied yourself with bait for several more traps.

Trap and Snare Construction

Traps and snares crush, choke, hang, or entangle the prey. A single trap or snare will commonly incorporate two or more of these principles. The mechanisms that provide power to the trap are almost always very simple. The struggling victim, the force of gravity, or a bent sapling's tension provides the power.

The heart of any trap or snare is the trigger. When planning a trap or snare, ask yourself how it should affect the prey, what is the source of power, and what will be the most efficient trigger. Your answers will help you devise a specific trap for a specific species. Traps are designed to catch and hold or to catch and kill. Snares are traps that incorporate a noose to accomplish either function.

Simple Snare

A simple snare (Figure 2-3) consists of a noose placed over a trail or den hole and attached to a firmly planted stake. If the noose is some

Figure 2-3: Simple snare

type of cordage placed upright on a game trail, use small twigs or blades of grass to hold it up. Filaments from spider webs are excellent for holding nooses open. Make sure the noose is large enough to pass freely over the animal's head. As the animal continues to move, the noose tightens around its neck. The more the animal struggles, the tighter the noose gets. This type of snare usually does not kill the animal. If you use cordage, it may loosen enough to slip off the animal's neck. Wire is therefore the best choice for a simple snare.

Drag Noose
Use a drag noose on an animal run (Figure 2-4). Place forked sticks on either side of the run and lay a sturdy crossmember across them. Tie the noose to the crossmember and hang it at a height above the animal's head. (Nooses designed to catch by the head should never be low enough for the prey to step into with a foot.) As the noose tightens around the animal's neck, the animal pulls the crossmember from the forked sticks and drags it along. The surrounding vegetation quickly catches the crossmember and the animal becomes entangled.

Twitch-Up
A twitch-up is a supple sapling, which, when bent over and secured with a triggering device, will provide power to a variety of snares.

UNDER ↑ ↑ OVER

WIRE IS TWISTED
TO ITS ENDS

DOUBLE WIRE
LOCKING LOOP

FUNNELING

Figure 2-4: Drag noose

Select a hardwood sapling along the trail. A twitch-up will work much faster and with more force if you remove all the branches and foliage.

Twitch- Up Snare

A simple twitch-up snare uses two forked sticks, each with a long and short leg (Figure 2-5). Bend the twitch-up and mark the trail below it. Drive the long leg of one forked stick firmly into the ground at that point. Ensure the cut on the short leg of this stick is parallel to the ground. Tie the long leg of the remaining forked stick to a piece of cordage secured to the twitch-up. Cut the short leg so that it catches on the short leg of the other forked stick. Extend a noose over the trail. Set the trap by bending the twitch-up and engaging the short legs of the forked sticks. When an animal catches its head in the noose, it pulls the forked sticks apart, allowing the twitch-up to spring up and hang the prey.

Note: Do not use green sticks for the trigger. The sap that oozes out could glue them together.

Figure 2-5: Twitch-up snare

Squirrel Pole
A squirrel pole is a long pole placed against a tree in an area showing a lot of squirrel activity (Figure 2-6). Place several wire nooses along the top and sides of the pole so that a squirrel trying to go up or down the pole will have to pass through one or more of them. Position the nooses (5 to 6 centimeters in diameter) about 2.5 centimeters off the pole. Place the top and bottom wire nooses 45 centimeters from the top and bottom of the pole to prevent the squirrel from getting its feet on a solid surface. If this happens, the squirrel will chew through the wire. Squirrels are naturally curious. After an initial period of caution, they will try to go up or down the pole and will get caught in a noose. The struggling animal will soon fall from the pole and strangle. Other squirrels will soon follow and, in this way, you can catch several squirrels. You can emplace multiple poles to increase the catch.

Ojibwa Bird Pole
An Ojibwa bird pole is a snare used by native Americans for centuries (Figure 2-7). To be effective, place it in a relatively open area away from tall trees. For best results, pick a spot near feeding areas, dusting areas, or watering holes. Cut a pole 1.8 to 2.1 meters long and trim away all limbs and foliage. Do not use resinous wood such as pine. Sharpen the upper end to a point, then drill a small diameter hole 5 to 7.5 centimeters down from the top. Cut a small stick 10 to

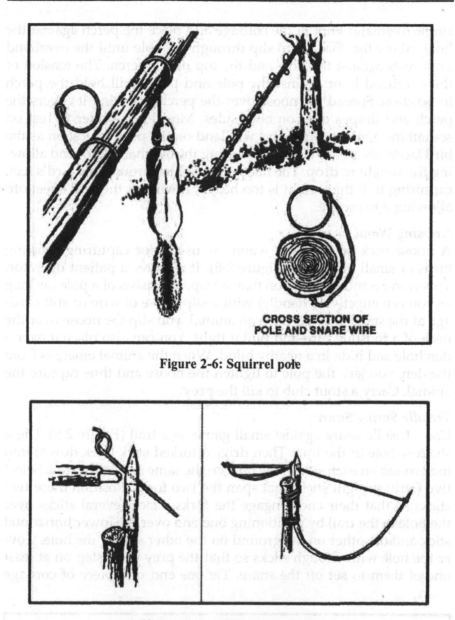

**CROSS SECTION OF
POLE AND SNARE WIRE**

Figure 2-6: Squirrel pole

Figure 2-7: Ojibwa bird pole

15 centimeters long and shape one end so that it will almost fit into the hole. This is the perch. Plant the long pole in the ground with the pointed end up. Tie a small weight, about equal to the weight of the targeted species, to a length of cordage. Pass the free end of the cordage through the hole, and tie a slip noose that covers the perch. Tie a

single overhand knot in the cordage and place the perch against the hole. Allow the cordage to slip through the hole until the overhand knot rests against the pole and the top of the perch. The tension of the overhand knot against the pole and perch will hold the perch in position. Spread the noose over the perch, ensuring it covers the perch and drapes over on both sides. Most birds prefer to rest on something above ground and will land on the perch. As soon as the bird lands, the perch will fall, releasing the overhand knot and allowing the weight to drop. The noose will tighten around the bird's feet, capturing it. If the weight is too heavy, it will cut the bird's feet off, allowing it to escape.

Noosing Wand

A noose stick or "noosing wand" is useful for capturing roosting birds or small mammals (Figure 2-8). It requires a patient operator. This wand is more a weapon than a trap. It consists of a pole (as long as you can effectively handle) with a slip noose of wire or stiff cordage at the small end. To catch an animal, you slip the noose over the neck of a roosting bird and pull it tight. You can also place it over a den hole and hide in a nearby blind. When the animal emerges from the den, you jerk the pole to tighten the noose and thus capture the animal. Carry a stout club to kill the prey.

Treadle Spring Snare

Use a treadle snare against small game on a trail (Figure 2-9). Dig a shallow hole in the trail. Then drive a forked stick (fork down) into the ground on each side of the hole on the same side of the trail. Select two fairly straight sticks that span the two forks. Position these two sticks so that their ends engage the forks. Place several sticks over the hole in the trail by positioning one end over the lower horizontal stick and the other on the ground on the other side of the hole. Cover the hole with enough sticks so that the prey must step on at least one of them to set off the snare. Tie one end of a piece of cordage

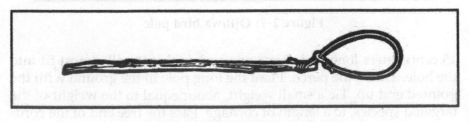

Figure 2-8: Noosing wand

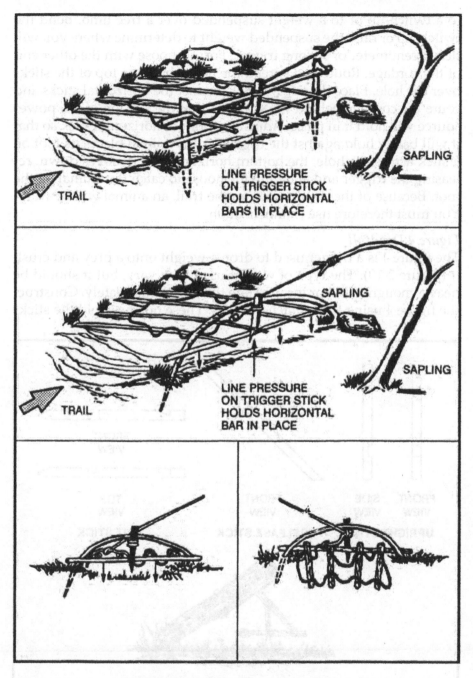

Figure 2-9: Treadle spring snare

to a twitch-up or to a weight suspended over a tree limb. Bend the twitch-up or raise the suspended weight to determine where you will tie a 5 centimeter or so long trigger. Form a noose with the other end of the cordage. Route and spread the noose over the top of the sticks over the hole. Place the trigger stick against the horizontal sticks and route the cordage behind the sticks so that the tension of the power source will hold it in place. Adjust the bottom horizontal stick so that it will barely hold against the trigger. A the animal places its foot on a stick across the hole, the bottom horizontal stick moves down, releasing the trigger and allowing the noose to catch the animal by the foot. Because of the disturbance on the trail, an animal will be wary. You must therefore use channelization.

Figure 4 Deadfall

The figure 4 is a trigger used to drop a weight onto a prey and crush it (Figure 2-10). The type of weight used may vary, but it should be heavy enough to kill or incapacitate the prey immediately. Construct the figure 4 using three notched sticks. These notches hold the sticks

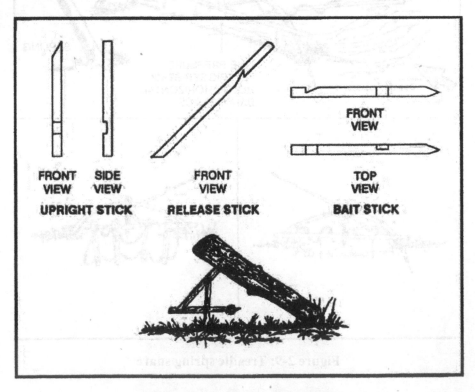

Figure 2-10: Figure 4 deadfall

together in a figure 4 pattern when under tension. Practice making this trigger beforehand; it requires close tolerances and precise angles in its construction.

Paiute Deadfall

The Paiute deadfall is similar to the figure 4 but uses a piece of cordage and a catch stick (Figure 2-11). It has the advantage of being easier to set than the figure 4. Tie one end of a piece of cordage to the lower end of the diagonal stick. Tie the other end of the cordage to another stick about 5 centimeters long. This 5-centimeter stick is the catch stick. Bring the cord halfway around the vertical stick with the catch stick at a 90-degree angle. Place the bait stick with one end against the drop weight, or a peg driven into the ground, and the other against the catch stick. When a prey disturbs the bait stick, it falls free, releasing the catch stick. As the diagonal stick flies up, the weight falls, crushing the prey.

Bow Trap

A bow trap is one of the deadliest traps. It is dangerous to man as well as animals (Figure 2-12). To construct this trap, build a bow and anchor it to the ground with pegs. Adjust the aiming point as you

Figure 2-11: Paiute deadfall

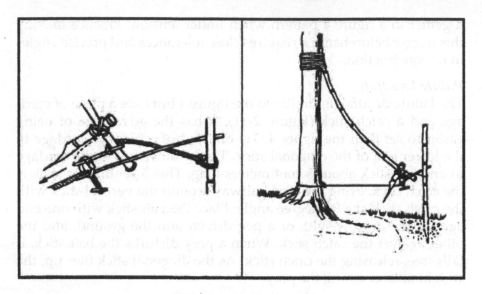

Figure 2-12: Bow trap

anchor the bow. Lash a toggle stick to the trigger stick. Two upright sticks driven into the ground hold the trigger stick in place at a point where the toggle stick will engage the pulled bow string. Place a catch stick between the toggle stick and a stake driven into the ground. Tie a tripwire or cordage to the catch stick and route it around stakes and across the game trail where you tie it off (as in Figure 2-12). When the prey trips the trip wire, the bow looses an arrow into it. A notch in the bow serves to help aim the arrow.

WARNING
This is a lethal trap. Approach it with caution and from the rear only!

Pig Spear Shaft

To construct the pig spear shaft, select a stout pole about 2.5 meters long (Figure 2-13). At the smaller end, firmly lash several small stakes. Lash the large end tightly to a tree along the game trail. Tie a length of cordage to another tree across the trail. Tie a sturdy, smooth stick to the other end of the cord. From the first tree, tie a trip wire or cord low to the ground, stretch it across the trail, and tie it to a catch

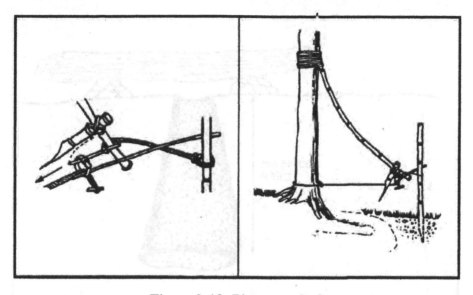

Figure 2-13: Pig spear shaft

stick. Make a slip ring from vines or other suitable material. Encircle the trip wire and the smooth stick with the slip ring. Emplace one end of another smooth stick within the slip ring and its other end against the second tree. Pull the smaller end of the spear shaft across the trail and position it between the short cord and the smooth stick. As the animal trips the tripwire, the catch stick pulls the slip ring off the smooth sticks, releasing the spear shaft that springs across the trail and impales the prey against the tree.

WARNING
This is a lethal trap. Approach it with caution!

Bottle Trap
A bottle trap is a simple trap for mice and voles (Figure 2-14). Dig a hole 30 to 45 centimeters deep that is wider at the bottom than at the top. Make the top of the hole as small as possible. Place a piece of bark or wood over the hole with small stones under it to hold it up 2.5 to 5 centimeters off the ground. Mice or voles will hide under the cover to escape danger and fall into the hole. They cannot climb out because of the wall's backward slope. Use caution when checking this trap; it is an excellent hiding place for snakes.

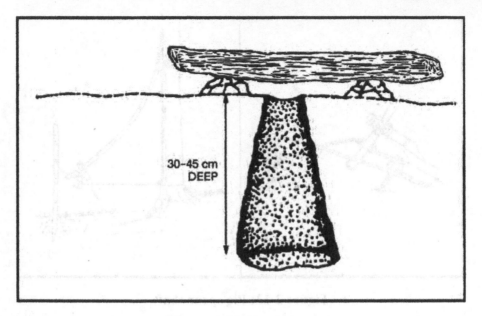

Figure 2-14: Bottle trap

KILLING DEVICES

There are several killing devices that you can construct to help you obtain small game to help you survive. The rabbit stick, the spear, the bow and arrow, and the sling are such devices.

Rabbit Stick

One of the simplest and most effective killing devices is a stout stick as long as your arm, from fingertip to shoulder, called a "rabbit stick." You can throw it either overhand or sidearm and with considerable force. It is very effective against small game that stops and freezes as a defense.

Spear

You can make a spear to kill small game and to fish. Jab with the spear, do not throw it. See Figure 2-20.

Bow and Arrow

A good bow is the result of many hours of work. You can construct a suitable short-term bow fairly easily. When it loses its spring or breaks, you can replace it. Select a hardwood stick about one meter long that is free of knots or limbs. Carefully scrape the large end down until it has the same pull as the small end. Careful examination will show the natural curve of the stick. Always scrape from the side that faces you, or the bow will break the first time you pull it. Dead, dry wood is preferable to green wood. To increase the pull, lash a second bow to the first, front to front, forming an "X" when viewed from the side. Attach the tips of the bows with cordage and only use a bowstring on one bow.

Select arrows from the straightest dry sticks available. The arrows should be about half as long as the bow. Scrape each shaft smooth all around. You will probably have to straighten the shaft. You can bend an arrow straight by heating the shaft over hot coals. Do not allow the shaft to scorch or burn. Hold the shaft straight until it cools.

You can make arrowheads from bone, glass, metal, or pieces of rock. You can also sharpen and fire harden the end of the shaft. To fire harden wood, hold it over hot coals, being careful not to bum or scorch the wood.

You must notch the ends of the arrows for the bowstring. Cut or file the notch; do not split it. Fletching (adding feathers to the notched end of an arrow) improves the arrow's flight characteristics, but is not necessary on a field-expedient arrow.

Sling

You can make a sling by tying two pieces of cordage, about sixty centimeters long, at opposite ends of a palm-sized piece of leather or cloth. Place a rock in the cloth and wrap one cord around the middle finger and hold in your palm. Hold the other cord between the forefinger and thumb. To throw the rock, spin the sling several times in a circle and release the cord between the thumb and forefinger. Practice to gain proficiency. The sling is very effective against small game. See Part V, Chapter 2 for more information about slings and arrows of outrageous fortune.

FISHING DEVICES

You can make your own fishhooks, nets and traps and use several methods to obtain fish in a survival situation.

Improvised Fishhooks

You can make field-expedient fishhooks from pins, needles, wire, small nails, or any piece of metal. You can also use wood, bone, coconut shell, thorns, flint, seashell, or tortoise shell. You can also make fish hooks from any combination of these items (Figure 2-15).

To make a wooden hook, cut a piece of hardwood about 2.5 centimeters long and about 6 millimeters in diameter to form the shank. Cut a notch in one end in which to place the point. Place the point

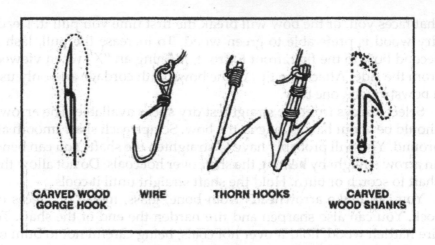

Figure 2-15: Improvised fishhooks

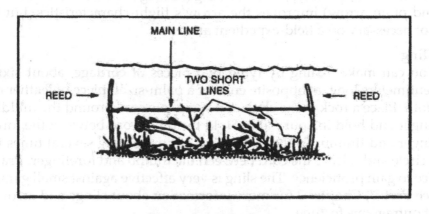

Figure 2-16: Stakeout

(piece of bone, wire, nail) in the notch. Hold the point in the notch and tie securely so that it does not move out of position. This is a fairly large hook. To make smaller hooks, use smaller material.

A gorge is a small shaft of wood, bone, metal, or other material. It is sharp on both ends and notched in the middle where you tie cordage. Bait the gorge by placing a piece of bait on it lengthwise. When the fish swallows the bait, it also swallows the gorge.

Stakeout
A stakeout is a fishing device you can use in a hostile environment (Figure 2-16). To construct a stakeout, drive two supple saplings into

the bottom of the lake, pond, or stream with their tops just below the water surface. Tie a cord between them and slightly below the surface. Tie two short cords with hooks or gorges to this cord, ensuring that they cannot wrap around the poles or each other. They should also not slip along the long cord. Bait the hooks or gorges.

Gill Net

If a gill net is not available, you can make one using parachute suspension line or similar material (Figure 2-17). Remove the core lines from the suspension line and tie the easing between two trees. Attach several core lines to the easing by doubling them over and tying them with prusik knots or girth hitches. The length of the desired net and the size of the mesh determine the number of core lines used and the space between them. Starting at one end of the easing, tie the second and the third core lines together using an overhand knot. Then tie the fourth and fifth, sixth and seventh, and so on, until you reach the last core line. You should now have all core lines tied in pairs with a single core line hanging at each end. Start the second row with the first core line, tie it to the second, the third to the fourth, and so on.

To keep the rows even and to regulate the size of the mesh, tie a guideline to the trees. Position the guideline on the opposite side of

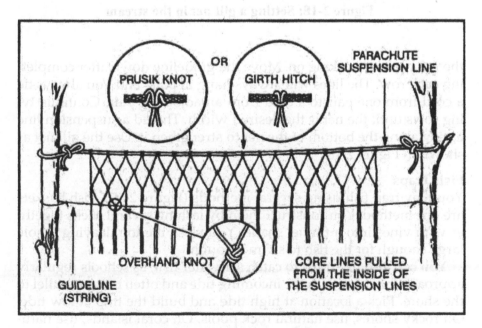

Figure 2-17: Making a gill net

Figure 2-18: Setting a gill net in the stream

the net you are working on. Move the guideline down after completing each row. The lines will always hang in pairs and you always tie a cord from one pair to a cord from an adjoining pair. Continue tying rows until the net is the desired width. Thread a suspension line easing along the bottom of the net to strengthen it. Use the gill net as shown in Figure 2-18.

Fish Traps
You may trap fish using several methods (Figure 2-19). Fish baskets are one method. You construct them by lashing several sticks together with vines into a funnel shape. You close the top, leaving a hole large enough for the fish to swim through.

You can also use traps to catch saltwater fish, as schools regularly approach the shore with the incoming tide and often move parallel to the shore. Pick a location at high tide and build the trap at low tide. On rocky shores, use natural rock pools. On coral islands, use natural pools on the surface of reefs by blocking the openings as the tide

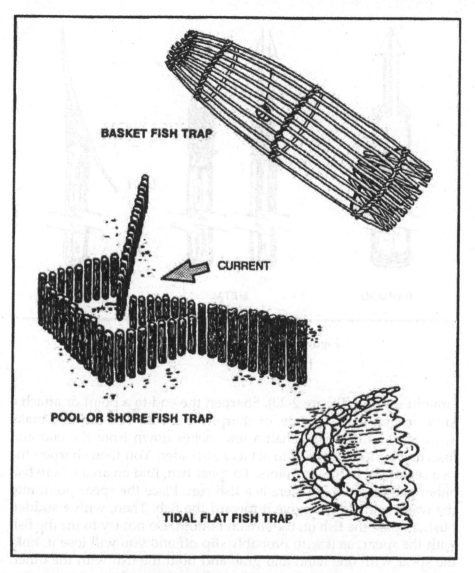

BASKET FISH TRAP

CURRENT

POOL OR SHORE FISH TRAP

TIDAL FLAT FISH TRAP

Figure 2-19: Various types of fish traps

recedes. On sandy shores, use sandbars and the ditches they enclose. Build the trap as a low stone wall extending outward into the water and forming an angle with the shore.

Spearfishing
If you are near shallow water (about waist deep) where the fish are large and plentiful, you can spear them. To make a spear, cut a long,

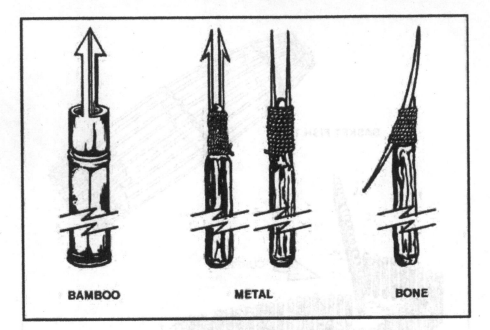

Figure 2-20: Types of spear points

straight sapling (Figure 2-20). Sharpen the end to a point or attach a knife, jagged piece of bone, or sharpened metal. You can also make a spear by splitting the shaft a few inches down from the end and inserting a piece of wood to act as a spreader. You then sharpen the two separated halves to points. To spear fish, find an area where fish either gather or where there is a fish run. Place the spear point into the water and slowly move it toward the fish. Then, with a sudden push, impale the fish on the stream bottom. Do not try to lift the fish with the spear, as it with probably slip off and you will lose it; hold the spear with one hand and grab and hold the fish with the other. Do not throw the spear, especially if the point is a knife. You cannot afford to lose a knife in a survival situation. Be alert to the problems caused by light refraction when looking at objects in the water.

Chop Fishing
At night, in an area with a good fish density, you can use a light to attract fish. Then, armed with a machete or similar weapon, you can gather fish using the back side of the blade to strike them. Do not use the sharp side as you will cut them in two pieces and end up losing some of the fish.

Fish Poison

Another way to catch fish is by using poison. Poison works quickly. It allows you to remain concealed while it takes effect. It also enables you to catch several fish at one time. When using fish poison, be sure to gather all of the affected fish, because many dead fish floating downstream could arouse suspicion. Some plants that grow in warm regions of the world contain rotenone, a substance that stuns or kills cold-blooded animals but does not harm persons who eat the animals. The best place to use rotenone, or rotenone-producing plants, is in ponds or the headwaiters of small streams containing fish. Rotenone works quickly on fish in water 21 degrees C (70 degrees F) or above. The fish rise helplessly to the surface. It works slowly in water 10 to 21 degrees C (50 to 70 degrees F) and is ineffective in water below 10 degrees C (50 degrees F). The plants in Figure 2-21, used as indicated, will stun or kill fish:

- *Anamirta cocculus:* This woody vine grows in southern Asia and on islands of the South Pacific. Crush the bean-shaped seeds and throw them in the water.
- *Croton tiglium:* This shrub or small tree grows in waste areas on islands of the South Pacific. It bears seeds in three angled capsules. Crush the seeds and throw them into the water.
- *Barringtonia:* These large trees grow near the sea in Malaya and parts of Polynesia. They bear a fleshy one-seeded fruit. Crush the seeds and bark and throw into the water.
- *Derris eliptica:* This large genus of tropical shrubs and woody vines is the main source of commercially produced rotenone. Grind the roots into a powder and mix with water. Throw a large quantity of the mixture into the water.
- *Duboisia:* This shrub grows in Australia and bears white clusters of flowers and berrylike fruit. Crush the plants and throw them into the water.
- *Tephrosia:* This species of small shrubs, which bears beanlike pods, grows throughout the tropics. Crush or bruise bundles of leaves and stems and throw them into the water.
- *Lime:* You can get lime from commercial sources and in agricultural areas that use large quantities of it. You may produce your own by burning coral or seashells. Throw the lime into the water.
- *Nut husks:* Crush green husks from butternuts or black walnuts. Throw the husks into the water.

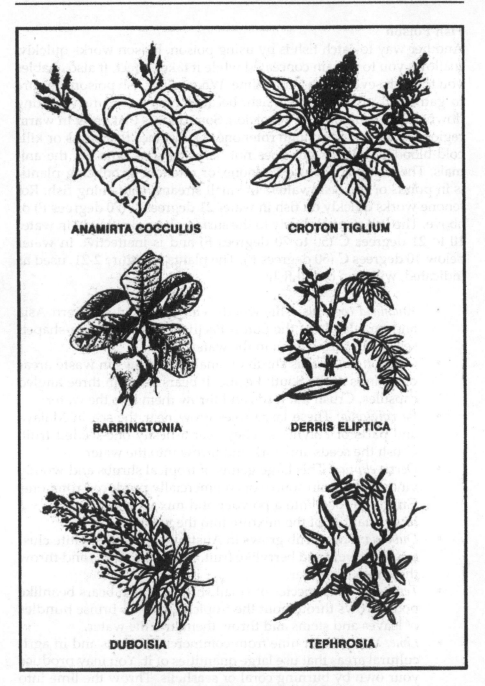

Figure 2-21: Fish-poisoning plants

PREPARATION OF FISH AND GAME FOR COOKING AND STORAGE

You must know how to prepare fish and game for cooking and storage in a survival situation. Improper cleaning or storage can result in inedible fish or game.

Fish

Do not eat fish that appears spoiled. Cooking does not ensure that spoiled fish will be edible. Signs of spoilage are—

- Sunken eyes.
- Peculiar odor.
- Suspicious color. (Gills should be red to pink. Scales should be a pronounced shade of gray, not faded.)
- Dents stay in the fish's flesh after pressing it with your thumb.
- Slimy, rather than moist or wet body.
- Sharp or peppery taste.

Eating spoiled or rotten fish may cause diarrhea, nausea, cramps, vomiting, itching, paralysis, or a metallic taste in the mouth. These symptoms appear suddenly, one to six hours after eating. Induce vomiting if symptoms appear.

Fish spoils quickly after death, especially on a hot day. Prepare fish for eating as soon as possible after catching it. Cut out the gills and large blood vessels that lie near the spine. Gut fish that is more than 10 centimeters long. Scale or skin the fish.

You can impale a whole fish on a stick and cook it over an open fire. However, boiling the fish with the skin on is the best way to get the most food value. The fats and oil are under the skin and, by boiling, you can save the juices for broth. You can use any of the methods used to cook plant food to cook fish. Pack fish into a ball of clay and bury it in the coals of a fire until the clay hardens. Break open the clay ball to get to the cooked fish. Fish is done when the meat flakes off. If you plan to keep the fish for later, smoke or fry it. To prepare fish for smoking, cut off the head and remove the backbone.

Snakes

To skin a snake, first cut off its head and bury it. Then cut the skin down the body 15 to 20 centimeters (Figure 2-22). Peel the skin back, then grasp the skin in one hand and the body in the other and pull apart. On large, bulky snakes it may be necessary to slit the belly skin.

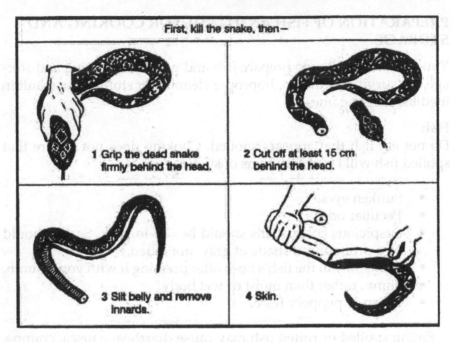

Figure 2-22: Cleaning a snake

Cook snakes in the same manner as small game. Remove the entrails and discard. Cut the snake into small sections and boil or roast it.

Birds
After killing the bird, remove its feathers by either plucking or skinning. Remember, skinning removes some of the food value. Open up the body cavity and remove its entrails, saving the craw (in seed-eating birds), heart, and liver. Cut off the feet. Cook by boiling or roasting over a spit. Before cooking scavenger birds, boil them at least 20 minutes to kill parasites.

Skinning and Butchering Game
Bleed the animal by cutting its throat. If possible, clean the carcass near a stream. Place the carcass belly up and split the hide from throat to tail, cutting around all sexual organs (Figure 2-23). Remove the musk glands at points A and B to avoid tainting the meat. For smaller mammals, cut the hide around the body and insert two fingers under the hide on both sides of the cut and pull both pieces off (Figure 2-24).

Note: when cutting the hide, insert the knife blade under the skin and turn the blade up so that only the hide gets cut. This will also prevent cutting hair and getting it on the meat.

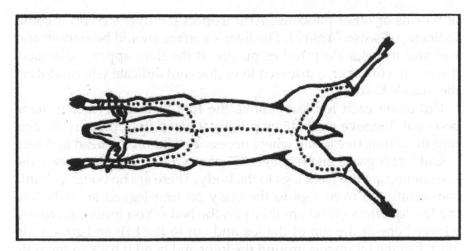

Figure 2-23: Skinning and butchering large game

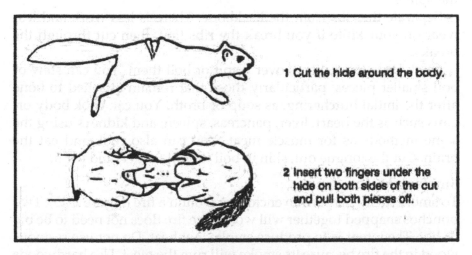

1 Cut the hide around the body.

2 Insert two fingers under the hide on both sides of the cut and pull both pieces off.

Figure 2-24: Skinning small game

Remove the entrails from smaller game by splitting the body open and pulling them out with the fingers. Do not forget the chest cavity. For larger game, cut the gullet away from the diaphragm. Roll the entrails out of the body. Cut around the anus, then reach into the lower abdominal cavity, grasp the lower intestine, and pull to remove. Remove the urine bladder by pinching it off and cutting it below the fingers. If you spill urine on the meat, wash it to avoid tainting the meat. Save the heart and liver. Cut these open and inspect for signs of worms or other parasites. Also inspect the liver's color; it could indicate a diseased animal. The liver's surface should be smooth and wet and its color deep red or purple. If the liver appears diseased, discard it. However, a diseased liver does not indicate you cannot eat the muscle tissue.

Cut along each leg from above the foot to the previously made body cut. Remove the hide by pulling it away from the carcass, cutting the connective tissue where necessary. Cut off the head and feet.

Cut larger game into manageable pieces. First, slice the muscle tissue connecting the front legs to the body. There are no bones or joints connecting the front legs to the body on four-legged animals. Cut the hindquarters off where they join the body. You must cut around a large bone at the top of the leg and cut to the ball and socket hip joint. Cut the ligaments around the joint and bend it back to separate it. Remove the large muscles (the tenderloin) that lie on either side of the spine.

Separate the ribs from the backbone. There is less work and less wear on your knife if you break the ribs first, then cut through the breaks.

Cook large meat pieces over a spit or boil them. You can stew or boil smaller pieces, particularly those that remain attached to bone after the initial butchering, as soup or broth. You can cook body organs such as the heart, liver, pancreas, spleen, and kidneys using the same methods as for muscle meat. You can also cook and eat the brain. Cut the tongue out, skin it, boil it until tender, and eat it.

Smoking Meat
To smoke meat, prepare an enclosure around a fire (Figure 2-25). Two ponchos snapped together will work. The fire does not need to be big or hot. The intent is to produce smoke, not heat. Do not use resinous wood in the fire because its smoke will ruin the meat. Use hardwoods to produce good smoke. The wood should be somewhat green. If it is too dry, soak it. Cut the meat into thin slices, no more than 6 centimeters thick, and drape them over a framework. Make sure none of the meat touches another piece. Keep the poncho enclosure around the meat to hold the smoke and keep a close watch on the fire. Do not let the fire get too hot. Meat smoked overnight in this manner will last about 1 week. Two days of continuous smoking will preserve the meat for 2 to 4 weeks. Properly smoked meat will look like a dark, curled, brittle stick and you can eat it without further cooking. You can also use a pit to smoke meat (Figure 2-26).

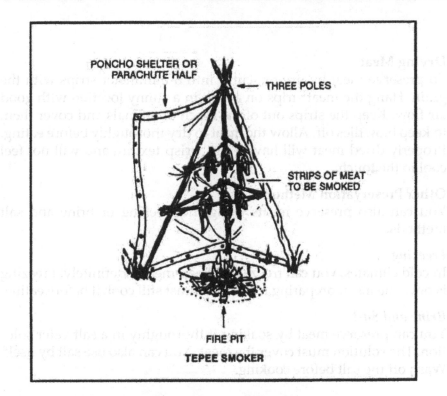

Figure 2-25: Smoking meat

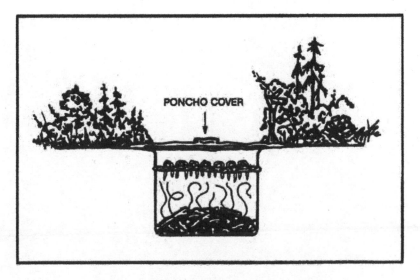

Figure 2-26: Smoking meat over a pit

Drying Meat

To preserve meat by drying, cut it into 6-millimeter strips with the grain. Hang the meat strips on a rack in a sunny location with good air flow. Keep the strips out of the reach of animals and cover them to keep blowflies off. Allow the meat to dry thoroughly before eating. Properly dried meat will have a dry, crisp texture and will not feel cool to the touch.

Other Preservation Methods

You can also preserve meats using the freezing or brine and salt methods.

Freezing

In cold climates, you can freeze and keep meat indefinitely. Freezing is not a means of preparing meat. You must still cook it before eating.

Brine and Salt

You can preserve meat by soaking it thoroughly in a saltwater solution. The solution must cover the meat. You can also use salt by itself. Wash off the salt before cooking.

CHAPTER 3

DANGEROUS INSECTS AND ARACHNIDS

Insects are often overlooked as a danger to the survivor. More people in the United States die each year from bee stings, and resulting anaphylactic shock, than from snake bites. A few other insects are venomous enough to kill, but often the greatest danger is the transmission of disease.

> ⚠ CAUTION
> Scorpions sting with their tails, causing local pain, swelling, possible incapacitation, and death.

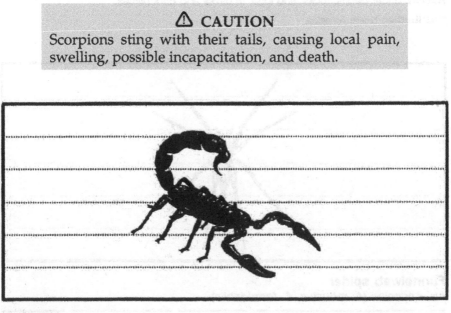

Scorpion
Scorpionidae order

Description: Dull brown, yellow, or black. Have 7.5- to 20-centimeter-long lobsterlike pincers and jointed tail usually held over the back. There are 800 species of scorpions.

Habitat: Decaying matter, under debris, logs, and rocks. Feeds at night. Sometimes hides in boots.

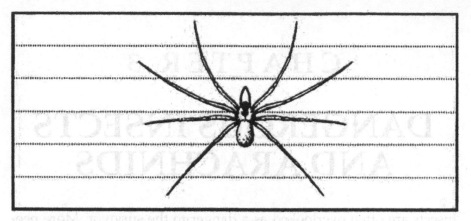

Brown house spider or brown recluse spider
Laxosceles reclusa

Description: Brown to black with obvious "fiddle" on back of head and thorax. Chunky body with long, slim legs 2.5 to 4 centimeters long.

Habitat: Under debris, rocks, and logs. In caves and dark places.

Distribution: North America.

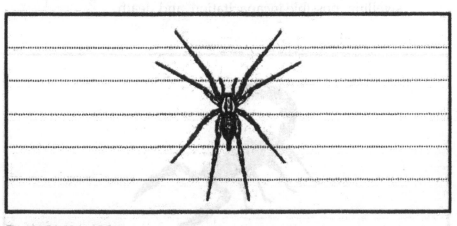

Funnelweb spider
Atrax species (*A. robustus, A. formidablis*)

Description: Large, brown, bulky spiders. Aggressive when disturbed.

Habitat: Woods, jungles, and brushy areas. Web has funnellike opening.

Distribution: Australia. (Other nonvenomous species worldwide.)

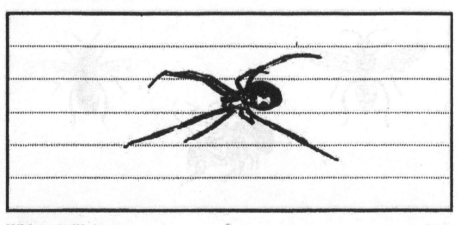

Tarantula
Theraphosidae and *Lycosa* species

Description: Very large, brown, black, reddish hairy spiders. Large fangs inflict painful bite.

Habitat: Desert areas, tropics.

Distribution: Americas, southern Europe.

Widow spider
Latrodectus species

Description: Dark spiders with light red or orange markings on female's abdomen.

Habitat: Under logs, rocks, and debris. In shaded places.

Distribution: Varied species worldwide. Black widow in United States, red widow in Middle East, and brown widow in Australia.

Note: Females are the poisonous gender. Red widow in the Middle East is the only spider known to be deadly to man.

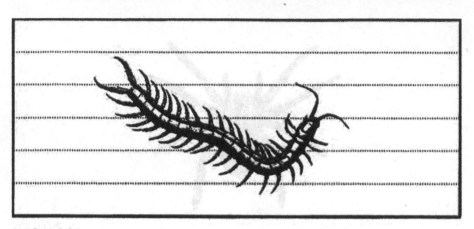

Centipede

Description: Multijointed body to 30 centimeters long. Dull orange to brown, with black point eyes at base of antennae. There are 2,800 species worldwide.

Habitat: Under bark and stones by day. Active at night.

Distribution: Worldwide.

Bee

Description: Insect with brown or black, thick, hairy bodies. Generally found in colonies. Many build wax combs.

Habitat: Hollow trees, caves, dwellings. Near water in desert areas.

Distribution: Worldwide.

Note: Bees have barbed stingers and die after stinging because their venom sac and internal organs are pulled out during the attack.

Tick

Description: Round body from size of pinhead to 2.5 centimeters. Has 8 legs and sucking mouth parts. There are 850 species worldwide.

Habitat: Mainly in forests and grasslands. Also in urban areas and farmlands.

Distribution: Worldwide.

CHAPTER 4

POISONOUS SNAKES AND LIZARDS

If you fear snakes, it is probably because you are unfamiliar with them or you have wrong information about them. There is no need for you to fear snakes if you know—

- Their habits.
- How to identify the dangerous kinds.
- Precautions to take to prevent snakebite.
- What actions to take in case of snakebite

For a man wearing shoes and trousers and living in a camp, the danger of being bitten by a poisonous snake is small compared to the hazards of malaria, cholera, dysentery, or other diseases.

Nearly all snakes avoid man if possible. Reportedly, however, a few—the king cobra of Southeast Asia, the bushmaster and tropical rattlesnake of South America, and the mamba of Africa—sometimes aggressively attack man, but even these snakes do so only occasionally. Most snakes get out of the way and are seldom seen.

WAYS TO AVOID SNAKEBITE

Snakes are widely distributed. They are found in all tropical, subtropical, and most temperate regions. Some species of snakes have specialized glands that contain a toxic venom and long hollow fangs to inject their venom.

Poisonous Snakes of the Americas

- American Copperhead (*Agkistrodon contortrix*)
- Bushmaster (*Lachesis mutus*)
- Coral snake (*Micrurus fulvius*)
- Cottonmouth (*Agkistrodon piscivorus*)

- Fer-de-lance (*Bothrops atrox*)
- Rattlesnake (*Crotalus* species)

Poisonous Snakes of Europe

- Common adder (*Vipers berus*)
- Pallas' viper (*Agkistrodon halys*)

Poisonous Snakes of Africa and Asia

- Boomslang (*Dispholidus typus*)
- Cobra (*Naja* species)
- Gaboon viper (*Bitis gabonica*)
- Green tree pit viper (*Trimeresurus gramineus*)
- Habu pit viper (*Trimeresurus flavoviridis*)
- Krait (*Bungarus caeruleus*)
- Malayan pit viper (*Callaselasma rhodostoma*)
- Mamba (*Dendraspis species*)
- Puff adder (*Bitis arietans*)
- Rhinoceros viper (*Bitis nasicornis*)
- Russell's viper (*Vipera russellii*)
- Sand viper (*Cerastes vipera*)
- Saw-scaled viper (*Echis carinatus*)
- Wagler's pit viper (*Trimeresurus wagleri*)

Poisonous Snakes of Australasia

- Death adder (*Acanthophis antarcticus*)
- Taipan (*Oxyuranus scutellatus*)
- Tiger snake (*Notechis scutatus*)
- Yellow-bellied sea snake (*Pelamis platurus*)

The polar regions are free of snakes due to their inhospitable environments. Other areas considered to be free of poisonous snakes are New Zealand, Cuba, Haiti, Jamaica, Puerto Rico, Ireland, Polynesia, and Hawaii.

There are no infallible rules for expedient identification of poisonous snakes in the field, because the guidelines all require close observation or manipulation of the snake's body. The best strategy is to leave all snakes alone. Where snakes are plentiful and poisonous species are present, the risk of their bites negates their food value.

Although venomous snakes use their venom to secure food, they also use it for self-defense. Human accidents occur when you don't see or hear the snake, when you step on them, or when you walk too close to them.

Follow these simple rules to reduce the chance of accidental snake-bite:

- Don't sleep next to brush, tall grass, large boulders, or trees. They provide hiding places for snakes. Place your sleeping bag in a clearing. Use mosquito netting tucked well under the bag. This netting should provide a good barrier.
- Don't put your hands into dark places, such as rock crevices, heavy brush, or hollow logs, without first investigating.
- Don't step over a fallen tree. Step on the log and look to see if there is a snake resting on the other side.
- Don't walk through heavy brush or tall grass without looking down. Look where you are walking.
- Don't pick up any snake unless you are absolutely positive it is not venomous.
- Don't pick up freshly killed snakes without first severing the head.
- The nervous system may still be active and a dead snake can deliver a bite.

SNAKE GROUPS

Snakes dangerous to man usually fall into two groups: proteroglypha and solenoglypha. Their fangs and their venom best describe these two groups (Table 4-1).

Fangs
The proteroglypha have, in front of the upper jaw and preceding the ordinary teeth, permanently erect fangs. These fangs are called fixed fangs.

Table 4-1: Snake group characteristics

Group	Fang Type	Venom Type
Proteroglypha	Fixed	Usually dominant neurotoxic
Solenoglypha	Folded	Usually dominant hemotoxic

The solenoglypha have erectile fangs; that is, fangs they can raise to an erect position. These fangs are called folded fangs.

Venom

The fixed-fang snakes (proteroglypha) usually have neurotoxic venoms. These venoms affect the nervous system, making the victim unable to breathe.

The folded-fang snakes (solenoglypha) usually have hemotoxic venoms, These venoms affect the circulatory system, destroying blood cells, damaging skin tissues, and causing internal hemorrhaging.

Remember, however, that most poisonous snakes have both neurotoxic and hemotoxic venom. Usually one type of venom in the snake is dominant and the other is weak.

Poisonous Versus Nonpoisonous Snakes

No single characteristic distinguishes a poisonous snake from a harmless one except the presence of poison fangs and glands. Only in dead specimens can you determine the presence of these fangs and glands without danger.

DESCRIPTIONS OF POISONOUS SNAKES

There are many different poisonous snakes throughout the world. It is unlikely you will see many except in a zoo. This manual describes only a few poisonous snakes. You should, however, be able to spot a poisonous snake if you–

- Learn about the two groups of snakes and the families in which they fall (Table 4-2).
- Examine the pictures and read the descriptions of snakes in this appendix.

Viperidae

The viperidae or true vipers usually have thick bodies and heads that are much wider than their necks (Figure 4-1). However, there are many different sizes, markings, and colorations.

This snake group has developed a highly sophisticated means for delivering venom. They have long, hollow fangs that perform like hypodermic needles. They deliver their venom deep into the wound.

The fangs of this group of snakes are movable. These snakes fold their fangs into the roof of their mouths. When they strike, their fangs

Table 4-2: Clinical effects of snake bites

Group	Family	Local Effects	Systemic Effects
Solenoglypha *Usually dominant hemotoxic venom affecting the circulatory system*	Viperidae True vipers with movable front fangs	Strong pain, swelling, necrosis	Hemorrhaging, internal organ breakdown, destroying of blood cells
	Crotalidae Pit vipers with movable front fangs		
	Trimeresurus		
Proteroglypha *Usually dominant neurotoxic venom affecting the nervous system*	Elapidae Fixed front fangs		
	Cobra	Various pains, swelling, necrosis	Respiratory collapse
	Krait	No local effects	Respiratory collapse
	Micrurus	Little or no pain; no local symptoms	Respiratory collapse
	Laticaudinae and Hydrophidae Ocean-living with fixed front fangs	Pain and local swelling	Respiratory collapse

Note: The venom of the Gaboon viper, the rhinoceros viper, the tropical rattlesnake, and the Mojave rattlesnake is both strongly hemotoxic and strongly neurotoxic.

come forward, stabbing the victim. The snake controls the movement of its fangs; fang movement is not automatic. The venom is usually hemotoxic. There are, however, several species that have large quantities of neurotoxic elements, thus making them even more dangerous. The vipers are responsible for many human fatalities around the world.

Crotalidae
The crotalids, or pit vipers (Figure 4-2), may be either slender or thick-bodied. Their heads are usually much wider than their necks.

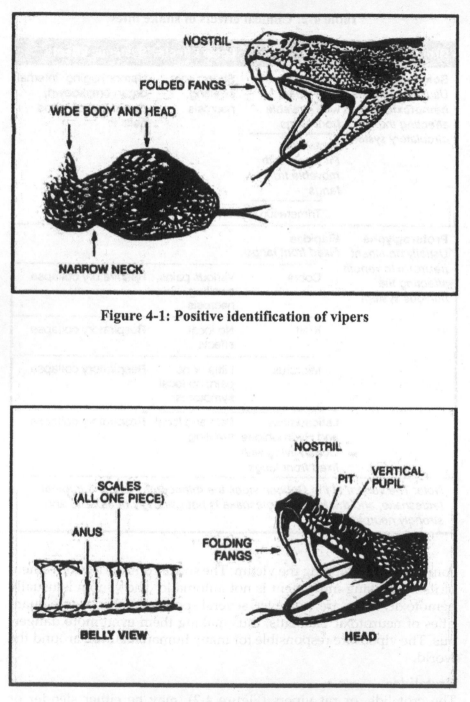

Figure 4-1: Positive identification of vipers

Figure 4-2: Positive identification of pit vipers

These snakes take their name from the deep pit located between the eye and the nostril. They are commonly brown with dark blotches, though some kinds are green.

Rattlesnakes, copperheads, cottonmouths, and several species of dangerous snakes from Central and South America, Asia, China, and India fall into the pit viper group. The pit is a highly sensitive organ capable of picking up the slightest temperature variance. Most pit vipers are nocturnal. They hunt for food at night with the aid of these specialized pits that let them locate prey in total darkness. Rattlesnakes are the only pit vipers that possess a rattle at the tip of the tail.

India has about 12 species of these snakes. You find them in trees or on the ground in all types of terrain. The tree snakes are slender; the ground snakes are heavy-bodied. All are dangerous.

China has a pit viper similar to the cottonmouth found in North America. You find it in the rocky areas of the remote mountains of South China. It reaches a length of 1.4 meters but is not vicious unless irritated. You can also find a small pit viper, about 45 centimeters long, on the plains of eastern China. It is too small to be dangerous to a man wearing shoes.

There are about 27 species of rattlesnakes in the United States and Mexico. They vary in color and may or may not have spots or blotches. Some are small while others, such as the diamondbacks, may grow to 2.5 meters long.

There are five kinds of rattlesnakes in Central and South America, but only the tropical rattlesnake is widely distributed. The rattle on the tip of the tail is sufficient identification for a rattlesnake.

Most will try to escape without a fight when approached, but there is always a chance one will strike at a passerby. They do not always give a warning; they may strike first and rattle afterwards or not at all.

The genus Trimeresurus is a subgroup of the crotalidae. These are Asian pit vipers. These pit vipers are normally tree-loving snakes with a few species living on the ground. They basically have the same characteristics of the crotalidae–slender build and very dangerous. Their bites usually are on the upper extremities–head, neck, and shoulders. Their venom is largely hemotoxic.

Elapidae

A group of highly dangerous snakes with powerful neurotoxic venom that affects the nervous system, causing respiratory paralysis. Included in this family are coral snakes, cobras, mambas, and all the Australian venomous snakes. The coral snake is small and has

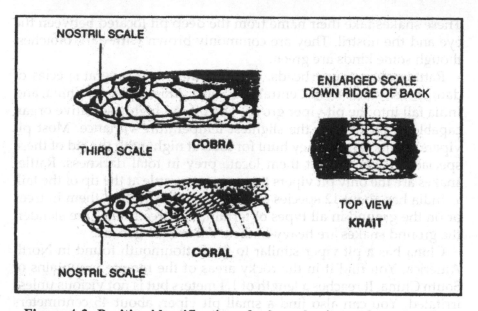

Figure 4-3: Positive identification of cobras, kraits, and coral snakes

caused human fatalities. The Australian death adder, tiger, taipan, and king brown snakes are among the most venomous in the world, causing many human fatalities.

Only by examining a dead snake can you positively determine if it is a cobra or a near relative (Figure 4-3). On cobras, kraits, and coral snakes, the third scale on the upper lip touches both the nostril scale and the eye. The krait also has a row of enlarged scales down its ridged back.

You can find the cobras of Africa and the Near East in almost any habitat. One kind may live in or near water, another in trees. Some are aggressive and savage. The distance a cobra can strike in a forward direction is equal to the distance its head is raised above the ground. Some cobras, however, can spit venom a distance of 3 to 3.5 meters. This venom is harmless unless it gets into your eyes; then it may cause blindness if not washed out immediately. Poking around in holes and rock piles is dangerous because of the chance of encountering a spitting cobra.

Laticaudinae and Hydrophidae
A subfamily of elapidae, these snakes are specialized in that they found a better environment in the oceans. Why they are in the oceans is not clear to science.

Sea snakes differ in appearance from other snakes in that they have an oarlike tail to aid in swimming. Some species of sea snakes have venom several times more toxic than the cobra's. Because of their marine environment, sea snakes seldom come in contact with humans. The exceptions are fishermen who capture these dangerous snakes in fish nets and scuba divers who swim in waters where sea snakes are found.

There are many species of sea snakes. They vary greatly in color and shape. Their scales distinguish them from eels that have no scales.

Sea snakes occur in salt water along the coasts throughout the Pacific. There are also sea snakes on the east coast of Africa and in the Persian Gulf. There are no sea snakes in the Atlantic Ocean.

There is no need to fear sea snakes. They have not been known to attack a man swimming. Fishermen occasionally get bit by a sea snake caught in a net. The bite is dangerous.

Colubridae
The largest group of snakes worldwide. In this family there are species that are rear-fanged; however, most are completely harmless to man. They have a venom-producing gland and enlarged, grooved rear fangs that allow venom to flow into the wound. The inefficient venom apparatus and the specialized venom is effective on cold-blooded animals (such as frogs and lizards) but not considered a threat to human life. The boomslang and the twig snake of Africa have, however, caused human deaths.

Table 4-3

Viperidae

- Common adder
- Long-nosed adder
- Gaboon viper
- Horned desert viper
- McMahon's viper
- Mole viper
- Palestinian viper
- Puff adder
- Rhinoceros viper
- Russell's viper
- Sand viper
- Saw-scaled viper
- Ursini's viper

Elapidae

- Australian copperhead
- Common cobra
- Coral snake
- Death adder
- Egyptian cobra
- Green mamba
- King cobra
- Krait
- Taipan
- Tiger snake

Crotallidae

- American copperhead
- Boomslang
- Bush viper
- Bushmaster
- Cottonmouth
- Easter diamondback rattlesnake
- Eyelash pit viper
- Fer-de-lance
- Green tre pit viper
- Habu pit viper
- Jumping ciper
- Malayan pit viper
- Mojave rattlesnake
- Pallas' viper
- Tropical rattlesnake
- Wagler's pit viper
- Western diamondback rattlesnake

Hydrophilidae

- Banded sea snake
- Yellow-bellied sea snake

LIZARDS

There is little to fear from lizards as long as you follow the same precautions as for avoiding snakebite. Usually, there are only two poisonous lizards: the Gila monster and the Mexican beaded lizard. The venom of both these lizards is neurotoxic. The two lizards are in the same family, and both are slow moving with a docile nature. The komodo dragon (Varanus komodoensis), although not poisonous, can be dangerous due to its large size. These lizards can reach lengths of 3 meters and weigh over 115 kilograms. Do not try to capture this lizard.

POISONOUS SNAKES OF THE AMERICAS
American copperhead

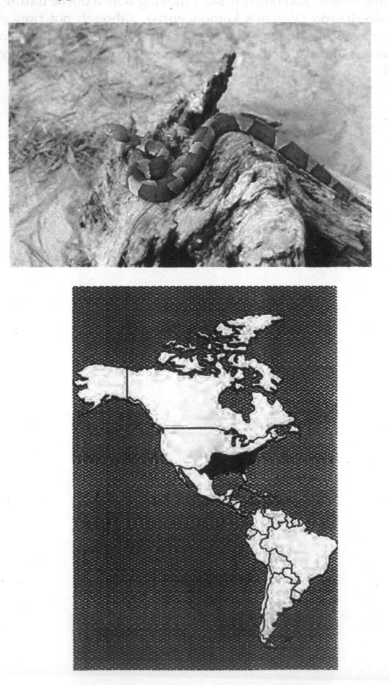

Agkistrodon contortrix

Description: Chestnut color dominates overall, with darker cross-bands of rich browns that become narrower on top and widen at the bottom. The top of the head is a coppery color.

Characteristic: Very common over much of its range, with a natural camouflage ability to blend in the environment. Copperheads are rather quiet and inoffensive in disposition but will defend themselves vigorously. Bites occur when the snakes are stepped on or when a victim is lying next to one. A copperhead lying on a bed of dead leaves becomes invisible. Its venom is hemotoxic.

Habitat: Found in wooded and rocky areas and mountainous regions.

Length: Average 60 centimeters, maximum 120 centimeters.

Distribution: Eastern Gulf States, Texas, Arkansas, Maryland, North Florida, Illinois, Oklahoma, Kansas, Ohio, New York, Alabama, Tennessee, and Massachusetts.

Bushmaster
Lachesis mutus

Description: The body hue is rather pale brown or pinkish, with a series of large bold dark brown or black blotches extending along the body. Its scales are extremely rough.

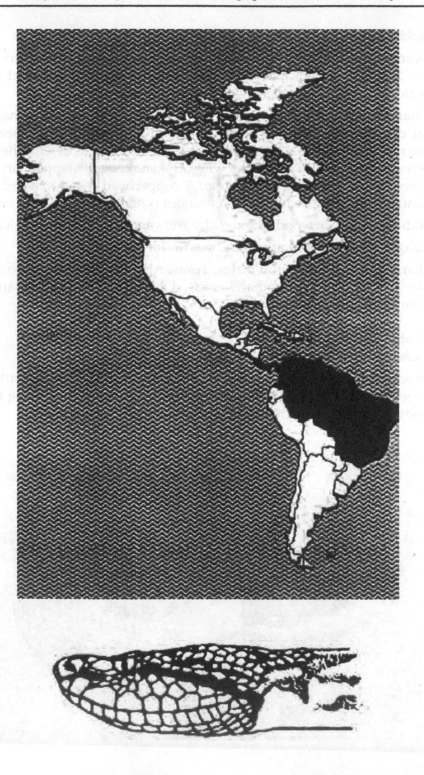

Characteristics: The World's largest pit viper has a bad reputation. This huge venomous snake is not common anywhere in its range. It lives in remote and isolated habitats and is largely nocturnal in its feeding habits; it seldom bites anyone, so few bites are recorded. A bite from one would indeed be very serious and fatal if medical aid was not immediately available. Usually, the bites occur in remote, dense jungles, many kilometers and several hours or even days away from medical help. Bushmaster fangs are long. In large bushmasters, they can measure 3.8 centimeters. Its venom is a powerful hemotoxin.

Habitat: Found chiefly in tropical forests in their range.

Length: Average 2.1 meters, maximum 3.7 meters.

Distribution: Nicaragua, Costa Rica, Panama, Trinidad, and Brazil.

Coral snake
Micrurus fulvius

Description: Beautifully marked with bright blacks, reds, and yellows. To identify the species, remember that when red touches yellow it is a coral snake.

Characteristics: Common over range, but secretive in its habits, therefore seldom seen. It has short fangs that are fixed in an erect position. It often chews to release its venom into a wound. Its venom is very powerful. The venom is neurotoxic, causing respiratory paralysis in the victim, who succumbs to suffocation.

Habitat: Found in a variety of habitats including wooded areas, swamps, palmetto and scrub areas. Coral snakes often venture into residential locations.

Length: Average 60 centimeters, maximum 115 centimeters.

Distribution: Southeast North Carolina, Gulf States, west central Mississippi, Florida, Florida Keys, and west to Texas. Another genus of coral snake is found in Arizona. Coral snakes are also found throughout Central and most South America.

Cottonmouth

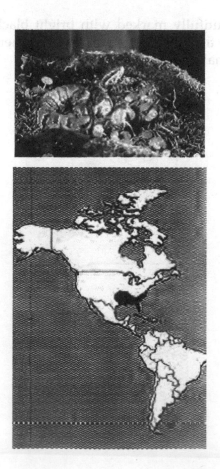

Agkistrodon piscivorus

Description: Colors are variable. Adults are uniformly olive brown or black. The young and subadults are strongly crossbanded with dark brown.

Characteristics: These dangerous semiaquatic snakes closely resemble harmless water snakes that have the same habitat. Therefore, it is best to leave all water snakes alone. Cottonmouths often stand their ground. An aroused cottonmouth will draw its head close to its body and open its mouth showing its white interior. Cottonmouth venom is hemotoxic and potent. Bites are prone to gangrene.

Habitat: Found in swamps, lakes, rivers, and ditches.

Length: Average 90 centimeters, maximum 1.8 meters.

Distribution: Southeast Virginia, west central Alabama, south Georgia, Illinois, east central Kentucky, south central Oklahoma, Texas, North Carolina, South Carolina, Florida, and the Florida Keys.

Eastern diamondback rattlesnake

Crotalus adamanteus

Description: Diamonds are dark brown or black, outlined by a row of cream or yellowish scales. Ground color is olive to brown.

Characteristics: The largest venomous snake in the United States. Large individual snakes can have fangs that measure 2.5 centimeters in a straight line. This species has a sullen disposition, ready to defend itself when threatened. Its venom is potent and hemotoxic, causing great pain and damage to tissue.

Habitat: Found in palmettos and scrubs, swamps, pine woods, and flatwoods. It has been observed swimming many miles out in the Gulf of Mexico, reaching some of the islands off the Florida coast.

Length: Average 1.4 meters, maximum 2.4 meters.

Distribution: Coastal areas of North Carolina, South Carolina, Louisiana, Florida, and the Florida Keys.

Eyelash pit viper

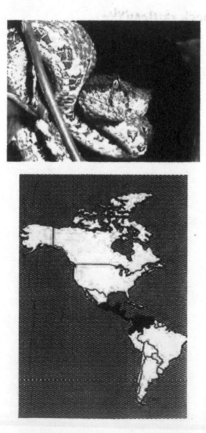

Bothrops schlegeli

Description: Identified by several spiny scales over each eye. Color is highly variable, from bright yellow over its entire body to reddish-yellow spots throughout the body.

Characteristics: Arboreal snake that seldom comes to the ground. It feels more secure in low-hanging trees where it looks for tree frogs and birds. It is a dangerous species because most of its bites occur on the upper extremities. It has an irritable disposition. It will strike with little provocation. Its venom is hemotoxic, causing severe tissue damage. Deaths have occurred from the bites of these snakes.

Habitat: Tree-loving species found in rain forests; common on plantations and in palm trees.

Length: Average 45 centimeters, maximum 75 centimeters.

Distribution: Southern Mexico, throughout Central America, Columbia, Ecuador, and Venezuela.

Fer-de-lance

Bothrops atrox

There are several closely related species in this group. All are very dangerous to man.

Description: Variable coloration, from gray to olive, brown, or reddish, with dark triangles edged with light scales. Triangles are narrow at the top and wide at the bottom.

Characteristics: This highly dangerous snake is responsible for a high mortality rate.
It has an irritable disposition, ready to strike with little provocation. The female fer-de-lance is highly prolific, producing up to 60 young born with a dangerous bite. The venom of this species is hemotoxic, painful, and hemorrhagic (causing profuse internal bleeding). The venom causes massive tissue destruction.

Habitat: Found on cultivated land and farms, often entering houses in search of rodents.

Length: Average 1.4 meters, maximum 2.4 meters.

Distribution: Southern Mexico, throughout Central and South America.

Jumping viper

Bothrops nummifer

Description: It has a stocky body. Its ground color varies from brown to gray and it has dark brown or black dorsal blotches. It has no pattern on its head.

Characteristics: It is chiefly a nocturnal snake. It comes out in the early evening hours to feed on lizards, rodents, and frogs. As the name implies, this species can strike with force as it actually leaves

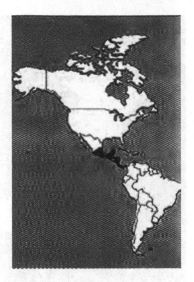

the ground. Its venom is hemotoxic. Humans have died from the bites inflicted by large jumping vipers. They often hide under fallen logs and piles of leaves and are difficult to see.

Habitat: Found in rain forests, on plantations, and on wooded hillsides.

Length: Average 60 centimeters, maximum 120 centimeters.

Distribution: Southern Mexico, Honduras, Guatemala, Costa Rica, Panama, and El Salvador.

Mojave rattlesnake

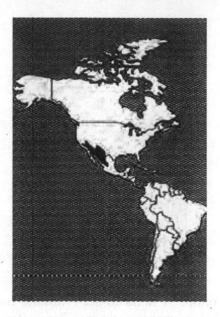

Crotalus scutulatus

Description: This snake's entire body is a pallid or sandy odor with darker diamond-shaped markings bordered by Lighter-colored scales and black bands around the tail.

Characteristics: Although this rattlesnake is of moderate size, its bite is very serious. Its venom has quantities of neurotoxic elements that affect the central nervous system. Deaths have resulted from this snake's bite.

Habitat: Found in arid regions, deserts, and rocky hillsides from sea level to 2400-meter elevations.

Length: Average 75 centimeters, maximum 1.2 meters.

Distribution: Mojave Desert in California, Nevada, southwest Arizona, and Texas into Mexico.

Tropical rattlesnake

Crotalus terrificus

Description: Coloration is light to dark brown with a series of darker rhombs or diamonds bordered by a buff color.

Characteristics: Extremely dangerous with an irritable disposition, ready to strike with little or no warning (use of its rattle). This species has a highly toxic venom containing neurotoxic and hemotoxic

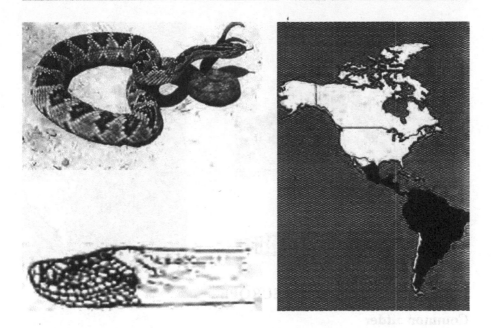

components that paralyze the central nervous system and cause great damage to tissue.

Habitat: Found in sandy places, plantations, and dry hillsides.

Length: Average 1.4 meters, maximum 2.1 meters.

Distribution: Southern Mexico, Central America, and Brazil to Argentina.

Western diamondback rattlesnake
Crotalus atrox

Description: The body is a light buff color with darker brown diamond-shaped markings. The tail has heavy black and white bands.

Characteristics: This bold rattlesnake holds its ground. When coiled and rattling, it is ready to defend itself. It injects a large amount of venom when it bites, making it one of the most dangerous snakes. Its venom is hemotoxic, causing considerable pain and tissue damage.

Habitat: It is a very common snake over its range. It is found in grasslands, deserts, woodlands, and canyons.

Length: Average 1.5 meters, maximum 2 meters.

Distribution: Southeast California, Oklahoma, Texas, New Mexico, and Arizona.

POISONOUS SNAKES OF EUROPE

Common adder
Vipera berus

Description: Its color is variable. Some adult specimens are completely black while others have a dark zigzag pattern running along the back.

Characteristics: The common adder is a small true viper that has a short temper and often strikes without hesitation. Its venom is

hemotoxic, destroying blood cells and causing tissue damage. Most injuries occur to campers, hikers, and field workers.

Habitat: Common adders are found in a variety of habitats, from grassy fields to rocky slopes, and on farms and cultivated lands.

Length: Average 45 centimeters, maximum 60 centimeters.

Distribution: Very common throughout most of Europe.

Long-nosed adder
Vipera ammodytes

Description: Coloration is gray, brown, or reddish with a dark brown or black zigzag pattern running the length of its back. A dark stripe is usually found behind each eye.

Characteristics: A small snake commonly found in much of its range. The term "long-nosed" comes from the projection of tiny scales located on the tip of its nose. This viper is responsible for many bites. Deaths have been recorded. Its venom is hemotoxic, causing severe pain and massive tissue damage. The rate of survival is good with medical aid.

Habitat: Open fields, cultivated lands, farms, and rocky slopes.

Length: Average 45 centimeters, maximum 90 centimeters.

Distribution: Italy, Yugoslavia, northern Albania, and Romania.

Pallas' viper
Agkistrodon halys

Description: Coloration is gray, tan, or yellow, with markings similar to those of the American copperhead.

Characteristics: This snake is timid and rarely strikes. Its venom is hemotoxic but rarely fatal.

Habitat: Found in open fields, hillsides, and farming regions.

Length: Average 45 centimeters, maximum 90 centimeters.

Distribution: Throughout southeastern Europe.

Ursini's viper
Vipera ursinii

Description: The common adder, long-nosed adder, and Ursini's viper basically have the same coloration and dorsal zigzag pattern. The exception among these adders is that the common adder and Ursini's viper lack the projection of tiny scales on the tip of the nose.

Characteristics: These little vipers have an irritable disposition. They will readily strike when approached. Their venom is hemotoxic. Although rare, deaths from the bites of these vipers have been recorded.

Habitat: Meadows, farmlands, rocky hillsides, and open, grassy fields.

Length: Average 45 centimeters, maximum 90 centimeters.

Distribution: Most of Europe, Greece, Germany, Yugoslavia, France, Italy, Hungary, Romania, Bulgaria, and Albania.

POISONOUS SNAKES OF AFRICA AND ASIA

Boomslang
Dispholidus typus

Description: Coloration varies but is generally green or brown, which makes it very hard to see in its habitat.

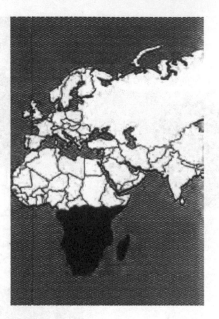

Characteristics: Will strike if molested. Its venom is hemotoxic; even small amounts cause severe hemorrhaging, making it dangerous to man.

Habitat: Found in forested areas. It will spend most of its time in trees or looking for chameleons and other prey in bushes.

Length: Generally less than 60 centimeters.

Distribution: Found throughout sub-Saharan Africa.

Bush viper

Atheris squamiger

Description: Often called leaf viper, its color varies from ground colors of pale green to olive, brown, or rusty brown. It uses its prehensile tail to secure itself to branches.

Characteristics: An arboreal species that often comes down to the ground to feed on small rodents. It is not aggressive, but it will defend itself when molested or touched. Its venom is hemotoxic; healthy adults rarely die from its bite.

Habitat: Found in rain forests and woodlands bordering swamps and forests. Often found in trees, low-hanging branches, or brush.

Length: Average 45 centimeters, maximum 75 centimeters.

Distribution: Most of Africa, Angola, Cameroon, Uganda, Kenya, and Zaire.

Common cobra
Naja naja

Description: Also known as the Asiatic cobra. Usually slate gray to brown overall. The back of the hood may or may not have a pattern.

Characteristics: A very common species responsible for many deaths each year. When aroused or threatened, the cobra will lift its head off the ground and spread its hood, making it more menacing. Its venom is highly neurotoxic, causing respiratory paralysis with some tissue

damage. The cobra would rather retreat if possible, but if escape is shut off, it will be a dangerous creature to deal with.

Habitat: Found in any habitat cultivated farms, swamps, open fields, and human dwelling where it searches for rodents.

Length: Average 1.2 meters, maximum 2.1 meters.

Distribution: All of Asia.

Egyptian cobra
Naja haje

Description: Yellowish, dark brown, or black uniform top with brown crossbands. Its head is sometimes black.

Characteristics: It is extremely dangerous. It is responsible for many human deaths. Once aroused or threatened, it will attack and continue the attack until it feels an escape is possible. Its venom is neurotoxic and much stronger than that of the common cobra. Its venom causes paralysis and death due to respiratory failure.

Habitat: Cultivated farmlands, open fields, and arid countrysides. It is often seen around homes searching for rodents.

Length: Average 1.5 meters, maximum 2.5 meters.

Distribution: Africa, Iraq, Syria, and Saudi Arabia.

Gaboon viper

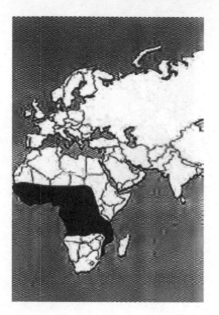

Bitis gabonica

Description: Pink to brown with a vertebral series of elongated yellowish or light brown spots connected by hourglass-shaped markings on each side. It has a dark brown stripe behind each eye. This dangerous viper is almost invisible on the forest floor. A 1.8-meter-long Gaboon viper could weigh 16 kilograms.

Characteristics: The largest and heaviest of all true vipers, having a very large triangular head. It comes out in the evening to feed. Fortunately, it is not aggressive, but it will stand its ground if approached. It bites when molested or stepped on. Its fangs are enormous, often measuring 5 centimeters long. It injects a large amount of venom when it strikes. Its venom is neurotoxic and hemotoxic.

Habitat: Dense rain forests. Occasionally found in open country.

Length: Average 1.2 meters, maximum 1.8 meters.

Distribution: Most of Africa.

Green mamba

Dendraspis angusticeps

Description: Most mambas are uniformly bright green over their entire body. The black mamba, the largest of the species, is uniformly olive to black.

Characteristics: The mamba is the dreaded snake species of Africa. Treat it with great respect. It is considered one of the most dangerous snakes known. Not only is it highly venomous but it is aggressive and its victim has little chance to escape from a bite. Its venom is highly neurotoxic.

Habitat: Mambas are at home in brush, trees, and low-hanging branches looking for birds, a usual diet for this species.

Length: Average 1.8 meters, maximum 3.7 meters.

Distribution: Most of Africa.

Green tree pit viper
Trimeresurus gramineus

Description: Uniform bright or dull green with light yellow on the facial lips.

Characteristics: A small arboreal snake of some importance, though not considered a deadly species. It is a dangerous species because most of its bites occur in the head, shoulder, and neck areas. It seldom comes to the ground. It feeds on young birds, lizards, and tree frogs.

Habitat: Found in dense rain forests and plantations.

Length: Average 45 centimeters, maximum 75 centimeters.

Distribution: India, Burma, Malaya, Thailand, Laos, Cambodia, Vietnam, China, Indonesia, and Formosa.

Habu pit viper
Trimeresurus flavoviridis

Description: Light brown or olive-yellow with black markings and a yellow or greenish-white belly.

Characteristics: This snake is responsible for biting many humans and its bite could be fatal. It is an irritable species ready to defend itself. Its venom is hemotoxic, causing pain and considerable tissue damage.

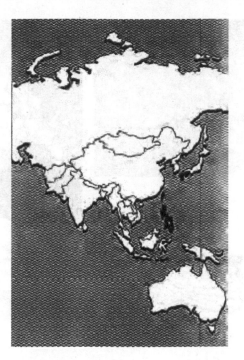

Habitat: Found in a variety of habitats, ranging from lowlands to mountainous regions. Often encountered in old houses and rock walls surroundings buildings.

Length: Average 1 meter, maximum 1.5 meters.

Distribution: Okinawa and neighboring islands and Kyushu.

Horned desert viper
Cerastes cerastes

Description: Pale buff color with obscure markings and a sharp spine (scale) over each eye.

Characteristics: As with all true vipers that live in the desert, it finds refuge by burrowing in the heat of the day, coming out at night to feed. It is difficult to detect when buried; therefore, many bites result from the snake being accidentally stepped on. Its venom is hemotoxic, causing severe damage to blood cells and tissue.

Habitat: Only found in very arid places within its range.

Length: Average 45 centimeters, maximum 75 centimeters.

Distribution: Arabian Peninsula, Africa, Iran, and Iraq.

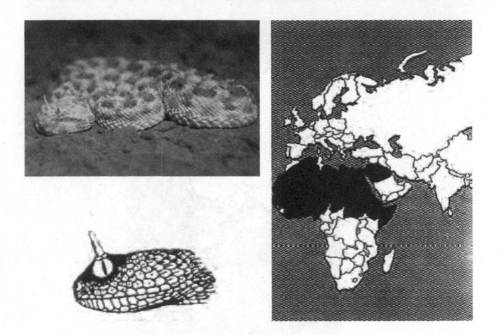

King cobra
Ophiophagus hannah

Description: Uniformly olive, brown, or green with ringlike cross-bands of black.

Characteristics: Although it is the largest venomous snake in the world and it has a disposition to go with this honor, it causes relatively few bites on humans. It appears to have a degree of intelligence. It avoids attacking another venomous snake for fear of being

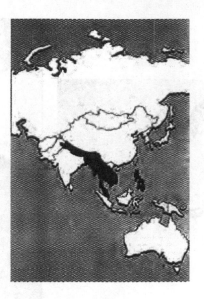

bitten. It feeds exclusively on harmless species. The female builds a nest then deposits her eggs. Lying close by, she guards the nest and is highly aggressive toward anything that closely approaches the nest. Its venom is a powerful neurotoxin. Without medical aid, death is certain for its victims.

Habitat: Dense jungle and cultivated fields.

Length: Average 3.5 meters, maximum 5.5 meters.

Distribution: Thailand, southern China, Malaysia Peninsula, and Phillipines.

Krait
Bungarus caeruleus

Description: Black or bluish-black with white narrow crossbands and a narrow head.

Characteristics: Kraits are found only in Asia. This snake is of special concern to man. It is deadly – about 15 times more deadly than the common cobra. It is active at night and relatively passive during the day. The native people often step on kraits while walking through their habitats. The krait has a tendency to seek shelter in sleeping bags, boots, and tents. Its venom is a powerful neurotoxin that causes respiratory failure.

Habitat: Open fields, human settlements, and dense jungle.

Length: Average 90 centimeters, maximum 1.5 meters.

Distribution: India, Sri Lanka, and Pakistan.

Levant viper

Vipera lebetina

Description: Gray to pale brown with large dark brown spots on the top of the black and a " ^ " mark on top of the head.

Characteristics: This viper belongs to a large group of true vipers. Like its cousins, it is large and dangerous. Its venom is hemotoxic. Many deaths have been reported from bites of this species. It is a strong snake with an irritable disposition; it hisses loudly when ready to strike.

Habitat: Varies greatly, from farmlands to mountainous areas.

Length: Average 1 meter, maximum 1.5 meters.

Distribution: Greece, Iraq, Syria, Lebanon, Turkey, Afganistan, lower portion of the former USSR, and Saudi Arabia.

Malayan pit viper
Callaselasma rhostoma

Description: Reddish running into pink tinge toward the belly with triangular-shaped, brown markings bordered with light-colored scales. The base of the triangular-shaped markings end at the midline. It has dark brown, arrow-shaped markings on the top and each side of its head.

Characteristics: This snake has long fangs, is ill-tempered, and is responsible for many bites. Its venom is hemotoxic, destroying blood cells and tissue, but a victim's chances of survival are good with medical aid. This viper is a ground dweller that moves into many

areas in search of food. The greatest danger is in stepping on the snake with bare feet.

Habitat: Rubber plantations, farms, rural villages, and rain forests.

Length: Average 60 centimeters, maximum 1 meter.

Distribution: Thailand, Laos, Cambodia, Java, Sumatra, Malaysia, Vietnam, Burma, and China.

McMahon's viper
Eristicophis macmahonii

Description: Sandy buff color dominates the body with darker brown spots on the side of the body. Its nose shield is broad, aiding in burrowing.

Characteristics: Very little is known about this species. It apparently is rare or seldom seen. This viper is very irritable; it hisses, coils, and strikes at any intruder that ventures too close. Its venom is highly hemotoxic, causing great pain and tissue damage.

Habitat: Arid or semidesert. It hides during the day's sun, coming out only at night to feed on rodents.

Length: Average 45 centimeters, maximum 1 meter.

Distribution: West Pakistan and Afghanistan.

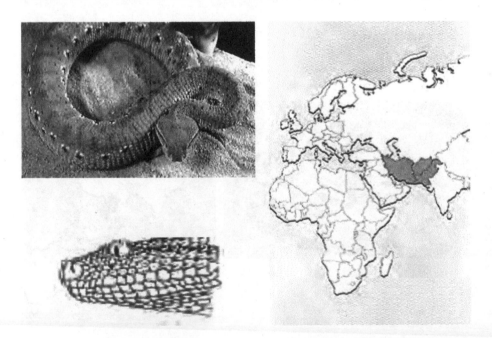

Mole viper or burrowing viper
Atracaspis microlepidota

Description: Uniformly black or dark brown with a small, narrow head.

Characteristics: A viper that does not look like one. It is small in size, and its small head does not indicate the presence of venom glands. It has a rather inoffensive disposition; however, it will quickly turn and bite if restrained or touched. Its venom is a potent hemotoxin for such a small snake. Its fangs are exceptionally long. A bite can result even when picking it up behind the head. It is best to leave this snake alone.

Habitat: Agricultural areas and arid localities

Length: Average 55 centimeters, maximum 75 centimeters

Distribution: Sudan, Ethiopia, Somaliland, Kenya, Tanganyika, Uganda, Cameroon, Niger, Congo, and Urundi.

Palestinian viper
Vipera palaestinae

Description: Olive to rusty brown with a dark V-shaped mark on the head and a brown, zigzag band along the back.

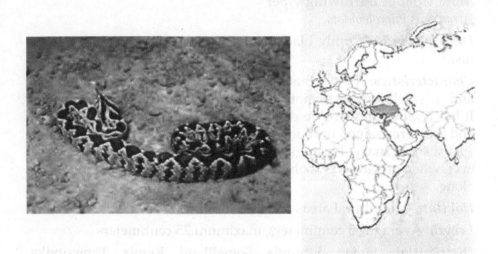

Characteristics: The Palestinian viper is closely related to the Russell's viper of Asia. Like its cousin, it is extremely dangerous. It is active and aggressive at night but fairly placid during the day. When threatened or molested, it will tighten its coils, hiss loudly, and strike quickly.

Habitat: Arid regions, but may be found around barns and stables. It has been seen entering houses in search of rodents.

Length: Average 0.8 meter, maximum 1.3 meters.

Distribution: Turkey, Syria, Palestine, Israel, Lebanon, and Jordan.

Puff adder

Bitis arietans

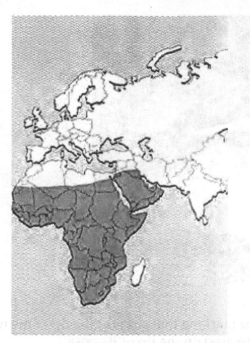

Description: Yellowish, light brown, or orange with chevron-shaped dark brown or black bars.

Characteristics: The puff adder is the second-largest of the dangerous vipers. It is one of the most common snakes in Africa. It is largely nocturnal, hunting at night and seeking shelter during the day's heat. It is not shy when approached. It draws its head close to its coils, makes a loud hissing sound, and is quick to strike any intruder. Its venom is strongly hemotoxic, destroying bloods cells and causingextensive tissue damage.

Habitat: And regions to swamps and dense forests. Common around human settlements.

Length: Average 12 meters, maximum 1.8 meters.

Distribution: Most of Africa, Saudi Arabia, Iraq, Lebanon, Israel, and Jordan.

Rhinoceros viper or river jack
Bitis nasicornis

Description: Brightly colored with purplish to reddish-brown markings and black and light olive markings along the back. On its head it

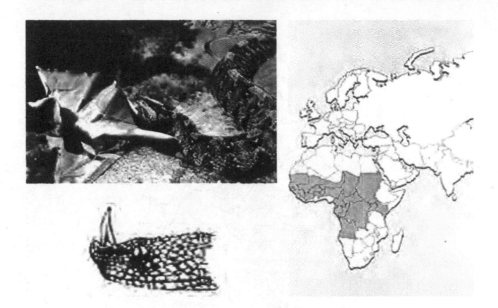

has a triangular marking that starts at the tip of the nose. It has a pair of long horns (scales) on the tip of its nose.

Characteristics: Its appearance is awesome; its horns and very rough scales give it a sinister look. It has an irritable disposition. It is not aggressive but will stand its ground ready to strike if disturbed. Its venom is neurotoxic and hemotoxic.

Habitat: Rain forests, along waterways, and in swamps.

Length: Average 75 centimeters, maximum 1 meter.

Distribution: Equatorial Africa.

Russell's viper

Vipera russellii

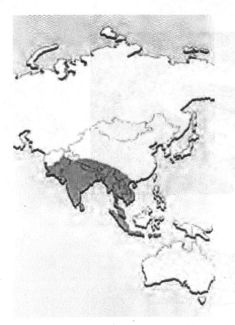

Description: Light brown body with three rows of dark brown or black splotches bordered with white or yellow extending its entire length.

Characteristics: This dangerous species is abundant over its entire range. It is responsible for more human fatalities than any other venomous snake. It is irritable. When threatened, it coils tightly, hisses, and strikes with such speed that its victim has little chance of escaping. Its hemotoxic venom is a powerful coagulant, damaging tissue and blood cells.

Habitat: Variable, from farmlands to dense rain forests. It is commonly found around human settlements.

Length: Average 1 meter, maximum 1.5 meters

Distribution: Sri Lanka, south China, India, Malaysian Peninsula, Java, Sumatra, Borneo, and surrounding islands.

Sand viper
Cerastes vipera

Description: Usually uniformly very pallid, with three rows of darker brown spots

Characteristics: A very small desert dweller that can bury itself in the sand during the day's heat. It is nocturnal, coming out at night

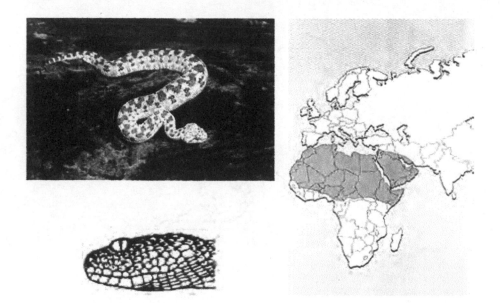

to feed on lizards and small desert rodents. It has a short temper and will strike several times. Its venom is hemotoxic.

Habitat: Restricted to desert areas.

Length: Average 45 centimeters, maximum 60 centimeters.

Distribution: Northern Sahara, Algeria, Egypt, Sudan, Nigeria, Chad, Somalia, and central Africa.

Saw-scaled viper
Echis carinatus

Description: Color is light buff with shades of brown, dull red, or gray. Its sides have a white or light-colored pattern. Its head usually has two dark stripes that start behind the eye and extend to the rear.

Characteristics: A small but extremely dangerous viper. It gets the name saw-scaled from rubbing the sides of its body together, producing a rasping sound. This ill-tempered snake will attack any intruder. Its venom is highly hemotoxic and quite potent. Many deaths are attributed to this species.

Habitat: Found in a variety of environments. It is common in rural settlements, cultivated fields, arid regions, barns, and rock walls.

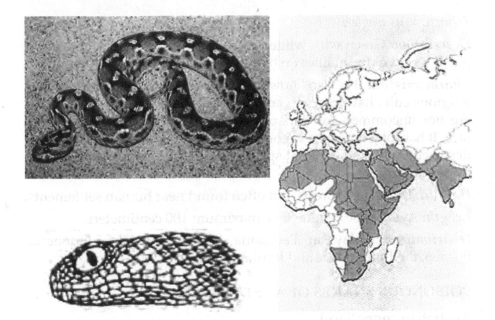

Length: Average 45 centimeters, maximum 60 centimeters.

Distribution: Asia, Syria, India, Africa, Iraq, Iran, Saudi Arabia, Pakistan, Jordan, Lebanon, Sri Lanka, Algeria, Egypt, and Israel.

Wagler's pit viper or temple viper

Trimeresurus wagleri

Description: Green with white crossbands edged with blue or purple. It has two dorsal lines on both sides of its head.

Characteristics: It is also known as the temple viper because certain religious cults have placed venomous snakes in their temples. Bites are not uncommon for the species; fortunately, fatalities are very rare. It has long fangs. Its venom is hemotoxic causing cell and tissue destruction. It is an arboreal species and its bites often occur on the upper extremities.

Habitat: Dense rain forests, but often found near human settlements.

Length: Average 60 centimeters, maximum 100 centimeters.

Distribution: Malaysian Peninsula and Archipelago, Indonesia, Borneo, the Philippines. and Ryuku Islands.

POISONOUS SNAKES OF AUSTRALASIA

Australian copperhead
Denisonia superba

Description: Coloration is reddish brown to dark brown. A few from Queensland are black.

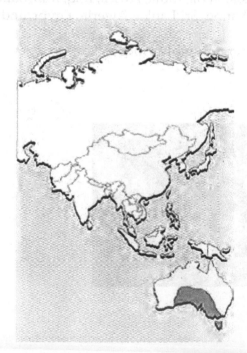

Characteristics: Rather sluggish disposition but will bite if stepped on. When angry, rears its head a few inches from the ground with its neck slightly arched. Its venom is neurotoxic.

Habitat: Swamps.

Length: Average 1.2 meters, maximum 1.8 meters.

Distribution: Tasmania, South Australia, Queensland, and Kangaroo Island.

Death adder
Acanthophis antarcticus

Description: Reddish, yellowish, or brown color with distinct dark brown crossbands. The end of its tail is black, ending in a hard spine.

Characteristics: When aroused, this highly dangerous snake will flatten its entire body, ready to strike over a short distance. It is nocturnal, hiding by day and coming out to feed at night. Although it has the appearance of a viper, it is related to the cobra family. Its venom is a powerful neurotoxin; it causes mortality in about 50 percent of the victims, even with treatment.

Habitat: Usually found in arid regions, fields, and wooded lands.

Length: Average 45 centimeters, maximum 90 centimeters.

Distribution: Australia, New Guinea, and Moluccas.

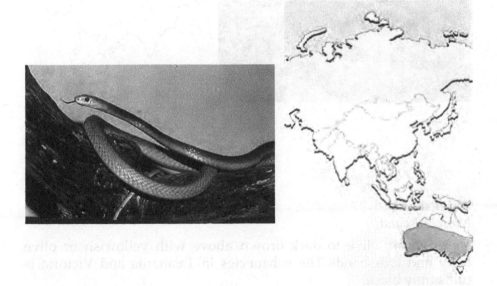

Taipan

Oxyuranus scutellatus

Description: Generally uniformly olive or dark brown, the head is somewhat darker brown.

Characteristics: Considered one of the most deadly snakes. It has an aggressive disposition. When aroused, it can display a fearsome appearance by flattening its head, raising it off the ground, waving it back and forth, and suddenly striking with such speed that the victim may receive several bites before it retreats. Its venom is a powerful neurotoxin, causing respiratory paralysis. Its victim has little chance for recovery without prompt medical aid.

Habitat: At home in a variety of habitats, it is found from the savanna forests to the inland plains.

Length: Average 1.8 meters, maximum 3.7 meters.

Distribution: Northern Australia and southern New Guinea.

Tiger snake

Notechis scutatus

Description: Olive to dark brown above with yellowish or olive belly and crossbands The subspecies in Tasmania and Victoria is uniformly black.

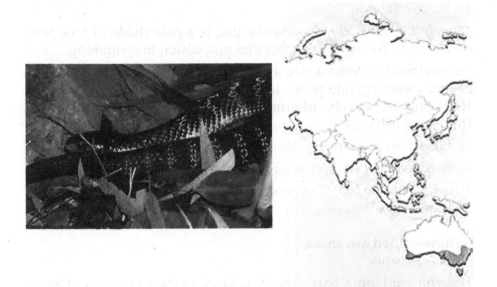

Characteristics: It is the most dangerous snake in Australia. It is very common and bites many humans. It has a very potent neurotoxic venom that attacks the nervous system. When aroused, it is aggressive and attacks any intruder. It flattens its neck making a narrow band.

Habitat: Found in many habitats from arid regions to human settlements along waterways to grasslands.

Length: Average 1.2 meters, maximum 1.8 meters.

Distribution: Australia, Tasmania, Bass Strait islands, and New Guinea.

POISONOUS SEA SNAKES

Banded sea snake

Laticauda colubrina

Description: Smooth-scaled snake that is a pale shade of blue with black bands. Its oarlike tail provides propulsion in swimming.

Characteristics: Most active at night, swimming close to shore and at times entering tide pools. Its venom is a very strong neurotoxin. Its victims are usually fishermen who untangle these deadly snakes from large fish nets.

Habitat: Common in all oceans, absent in the Atlantic Ocean.

Length: Average 75 centimeters, maximum 1.2 meters.

Distribution: Coastal waters of New Guinea, Pacific islands, the Philippines, Southeast Asia, Sri Lanka, and Japan.

Yellow-bellied sea snake
Pelamis platurus

Description: Upper part of body is black or dark brown and lower part is bright yellow.

Characteristics: A highly venomous snake belonging to the cobra family. This snake is truly of the pelagic species—it never leaves the water to come to shore. It has an oarlike tail to aid its swimming. This species is quick to defend itself. Sea snakes do not really strike, but

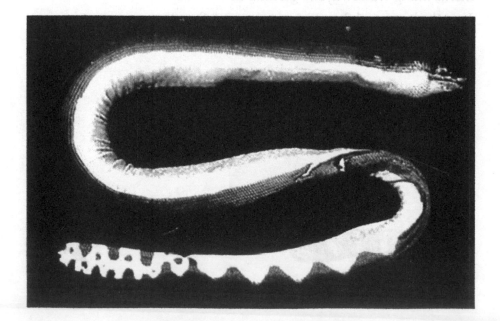

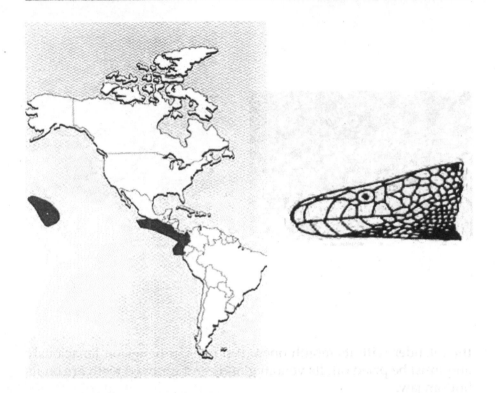

deliberately turn and bite if molested. A small amount of their neuro-toxic venom can cause death.

Habitat: Found in all oceans except the Atlantic Ocean.

Length: Average 0.7 meter, maximum 1.1 meters.

Distribution: Throughout the Pacific Ocean from many of the Pacific islands to Hawaii and to the coast of Costa Rica and Panama.

POISONOUS LIZARDS

Gila monster
Heloderma suspectum

Description: Robust, with a large head and a heavy tail. Its body is covered with beadlike scales. It is capable of storing fat against lean times when food is scarce. Its color is striking in rich blacks laced with yellow or pinkish scales.

Characteristics: Not an aggressive lizard, but it is ready to defend itself when provoked. If approached too closely, it will turn toward

the intruder with its mouth open. If it bites, it hangs on tenaciously and must be pried off. Its venom glands and grooved teeth are on its bottom jaw.

Habitat: Found in arid areas, coming out at night or early morning hours in search of small rodents and bird eggs. During the heat of the day it stays under brush or rocks.

Length: Average 30 centimeters, maximum 50 centimeters.

Distribution: Arizona, New Mexico, Utah, Nevada, northern Mexico, and extreme corner of southeast California.

Mexican beaded lizard
Heloderma horridum

Description: Less colorful than its cousin, the gila monster. It has black or pale yellow bands or is entirely black.

Characteristics: Very strong legs let this lizard crawl over rocks and dig burrows. It is short-tempered. It will turn and open its mouth in a threatening manner when molested. Its venom is hemotoxic and potentially dangerous to man.

Habitat: Found in arid or desert areas, often in rocky hillsides, coming out during evening and early morning hours.

Length: Average 60 centimeters, maximum 90 centimeters.
Distribution: Mexico through Central America.

CHAPTER 5

DANGEROUS FISH, MOLLUSKS, AND FRESHWATER ANIMALS

Since fish and mollusks may be one of your major sources of food, it is wise to know which ones are dangerous to you should you catch them. Know which ones are dangerous, what the dangers of the various fish are, what precautions to take, and what to do if you are injured by one of these fish.

Fish and mollusks will present a danger in one of three ways: by attacking and biting you, by injecting toxic venom into you through its venomous spines or tentacles, and through eating fish or mollusks whose flesh is toxic.

The danger of actually encountering one of these dangerous fish is relatively small, but it is still significant. Any one of these fish can kill you. Avoid them if at all possible.

DANGERS IN RIVERS

Common sense will tell you to avoid confrontations with hippopotami, alligators, crocodiles, and other large river creatures. There are, however, a few smaller river creatures with which you should be cautious.

Electric Eel
Electric eels (*Electrophorus electricus*) may reach 2 meters in length and 20 centimeters in diameter. Avoid them. They are capable of generating up to 500 volts of electricity in certain organs in their body. They use this shock to stun prey and enemies. Normally, you find these eels in the Orinoco and Amazon River systems in South America. They seem to prefer shallow waters that are more highly oxygenated and provide more food. They are bulkier than our native eels. Their upper body is dark gray or black, with a lighter-colored underbelly.

Piranha

Piranhas (*Serrasalmo* species) are another hazard of the Orinoco and Amazon River systems, as well as the Paraguay River Basin, where they are native. These fish vary greatly in size and coloration, but usually have a combination of orange undersides and dark tops. They have white, razor-sharp teeth that are clearly visible. They may be as long as 50 centimeters. Use great care when crossing waters where they live. Blood attracts them. They are most dangerous in shallow waters during the dry season.

Turtle

Be careful when handling and capturing large freshwater turtles, such as the snapping turtles and soft-shelled turtles of North America and the matamata and other turtles of South America. All of these turtles will bite in self-defense and can amputate fingers and toes.

Platypus

The platypus or duckbill (*Ornithorhyncus anatinus*) is the only member of its family and is easily recognized. It has a long body covered with grayish, short hair, a tail like a beaver, and a bill like a duck. Growing up to 60 centimeters in length, it may appear to be a good food source, but this egg-laying mammal, the only one in the world, is very dangerous. The male has a poisonous spur on each hind foot that can inflict intensely painful wounds. You find the platypus only in Australia, mainly along mud banks on waterways.

FISH THAT ATTACK MAN

The shark is usually the first fish that comes to mind when considering fish that attack man. Other fish also fall in this category, such as the barracuda, the moray eel, and the piranha.

Sharks

Whether you are in the water or in a boat or raft, you may see many types of sea life around you. Some may be more dangerous than others. Generally, sharks are the greatest danger to you. Other animals such as whales, porpoises, and stingrays may look dangerous, but really pose little threat in the open sea.

Of the many hundreds of shark species, only about 20 species are known to attack man. The most dangerous are the great white shark, the hammerhead, the make, and the tiger shark. Other sharks known to attack man include the gray, blue, lemon, sand, nurse, bull, and

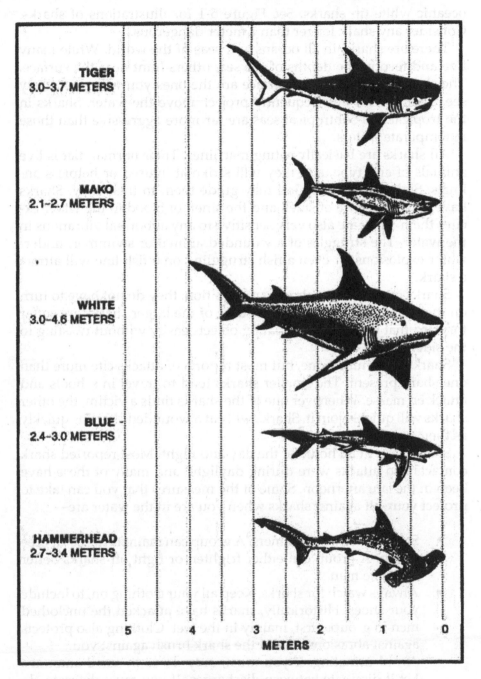

Figure 5-1: Sharks

oceanic white tip sharks. See Figure 5-1 for illustrations of sharks. Consider any shark longer than 1 meter dangerous.

There are sharks in all oceans and seas of the world. While many live and feed in the depths of the sea, others hunt near the surface. The sharks living near the surface are the ones you will most likely see. Their dorsal fins frequently project above the water. Sharks in the tropical and subtropical seas are far more aggressive than those in temperate waters.

All sharks are basically eating machines. Their normal diet is live animals of any type, and they will strike at injured or helpless animals. Sight, smell, or sound may guide them to their prey. Sharks have an acute sense of smell and the smell of blood in the water excites them. They are also very sensitive to any abnormal vibrations in the water. The struggles of a wounded animal or swimmer, underwater explosions, or even a fish struggling on a fish line will attract a shark.

Sharks can bite from almost any position; they do not have to turn on their side to bite. The jaws of some of the larger sharks are so far forward that they can bite floating objects easily without twisting to the side.

Sharks may hunt alone, but most reports of attacks cite more than one shark present. The smaller sharks tend to travel in schools and attack en masse. Whenever one of the sharks finds a victim, the other sharks will quickly join it. Sharks will eat a wounded shark as quickly as their prey.

Sharks feed at all hours of the day and night. Most reported shark contacts and attacks were during daylight, and many of these have been in the late afternoon. Some of the measures that you can take to protect yourself against sharks when you are in the water are—

- Stay with other swimmers. A group can maintain a 360-degree watch. A group can either frighten or fight off sharks better than one man.
- Always watch for sharks. Keep all your clothing on, to include your shoes. Historically, sharks have attacked the unclothed men in groups first, mainly in the feet. Clothing also protects against abrasions should the shark brush against you.
- Avoid urinating. If you must, only do so in small amounts. Let it dissipate between discharges. If you must defecate, do so in small amounts and throw it as far away from you as possible. Do the same if you must vomit.

If a shark attack is imminent while you are in the water, splash and yell just enough to keep the shark at bay. Sometimes yelling underwater or slapping the water repeatedly will scare the shark away. Conserve your strength for fighting in case the shark attacks.

If attacked, kick and strike the shark. Hit the shark on the gills or eyes if possible. If you hit the shark on the nose, you may injure your hand if it glances off and hits its teeth.

When you are in a raft and see sharks—

- Do not fish. If you have hooked a fish, let it go. Do not clean fish in the water.
- Do not throw garbage overboard.
- Do not let your arms, legs, or equipment hang in the water.
- Keep quiet and do not move around.
- Bury all dead as soon as possible. If there are many sharks in the area, conduct the burial at night.

When you are in a raft and a shark attack is imminent, hit the shark with anything you have, except your hands. You will do more damage to your hands than the shark. If you strike with an oar, be careful not to lose or break it.

If bitten by a shark, the most important measure for you to take is to stop the bleeding quickly. Blood in the water attracts sharks. Get yourself or the victim into a raft or to shore as soon as possible. If in the water, form a circle around the victim (if not alone), and stop the bleeding with a tourniquet.

Other Ferocious Fish
In salt water, other ferocious fish include the barracuda, sea bass, and moray eel (Figure 5-2). The sea bass is usually an open water fish. It is dangerous due to its large size. It can remove large pieces of flesh from a human. Barracudas and moray eels have been known to attack man and inflict vicious bites. Be careful of these two species when near reefs and in shallow water. Moray eels are very aggressive when disturbed.

VENOMOUS FISH AND INVERTEBRATES

There are several species of venomous fish and invertebrates, all of which live in salt water. All of these are capable of injecting poisonous venom through spines located in their fins, tentacles, or bites. Their venoms cause intense pain and are potentially fatal.

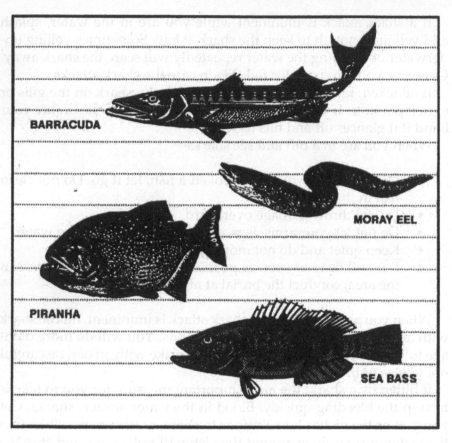

Figure 5-2: Ferocious fish

If injured by one of these fish or invertebrates, treat the injury as for snakebite.

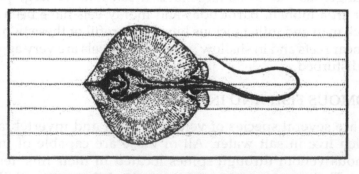

Figure 5-3: Stingray (Dasyatidae species)

Stingrays inhabit shallow water, especially in the tropics and in temperate regions as well. All have distinctive ray shape but coloration may make them hard to spot unless they are swimming. The venomous, barbed spines in their tails can cause severe or fatal injury. When moving about in shallow water, wear some form of footwear and shuffle your feet along the bottom, rather than picking up your feet and stepping.

Figure 5-4: Rabbitfish (Siganidae species)

Rabbitfish are found predominantly on reefs in the Pacific and Indian Oceans. They average about 30 centimeters long and have very sharp spines in their fins. The spines are venomous and can inflict intense pain. Rabbitfish are considered edible by native peoples where the fish are found, but deaths occur from careless handling. Seek other nonpoisonous fish to eat if possible.

Figure 5-5: Scorpion fish or Zebra fish (Scorpaenidae species)

Scorpion fish or Zebra fish live mainly in the reefs in the Pacific and Indian oceans, and occasionally in the Mediterranean and Aegean seas. They vary from 30 to 90 centimeters long, are unusually reddish in coloration, and have long, wavy fins and spines. They inflict an intensely painful sting.

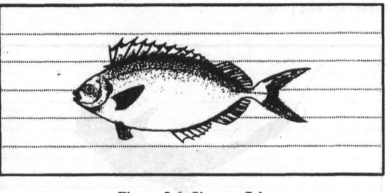

Figure 5-6: Siganus fish

The siganus fish is small, about 10 to 15 centimeters long, and looks much like a small tuna. It has venomous spines in its dorsal and ventral fins. These spines can inflict painful stings.

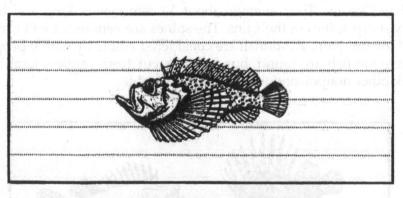

Figure 5-7: Stonefish (Synanceja species)

Stonefish are found in the tropical waters of the Pacific and Indian oceans. Averaging about 30 centimeters in length, their subdued colors and lumpy shape provide them with exceptional camouflage. When stepped on, the fins in the dorsal spine inflict an extremely painful and sometimes fatal wound.

Figure 5-8: Tang or surgeonfish (Acanthuridae species)

Tang or surgeonfish average 20 to 25 centimeters in length, with a deep body, small mouth, and bright coloration. They have scalpellike spines on the side of the tail that cause fatal and extremely painful wounds including infection, envenomation, and blood loss.

Figure 5-9: Toadfish (Batrachoidiae species)

Toadfish are found in the tropical waters of the Gulf Coast of the United States and along both coasts of South and Central America They are between 17.5 and 25 centimeters long and have a dull color and large mouths. They bury themselves in the sand and may be easily stepped on. They have very sharp, extremely poisonous spines on the dorsal fin (back).

Figure 5-10: Weever fish (Trachinidae species)

The weever fish is a tropical fish that is fairly slim and about 30 centimeters long. Its gills and all fins have venomous spines that cause a painful wound. They are found off the coasts of Europe, Africa, and the Mediterranean.

Figure 5-11: Blue-ringed octopus (Hapalochlaenalunulata)

This small octopus is usually found on the Great Barrier Reef off eastern Australia. It is grayish-white with iridescent blue ringlike markings. This octopus usually will not bite unless stepped on or handled. Its bite is extremely poisonous and frequently lethal.

Figure 5-12: Portuguese man-of-war (Physalis species)

Although it resembles a jellyfish, the Portuguese man-of-war is actually a colony of sea animals. Mainly found in tropical regions, the Gulf Stream current can carry it as far as Europe. It is also found as far south as Australia. The pink or purple floating portion of the man-of-war may be as small as 15 centimeters, but the tentacles can reach

12 meters in length. These tentacles inflict a painful and incapacitating sting, but the sting is rarely fatal. Avoid the tentacles of any jellyfish, even if washed up on the beach and apparently dead.

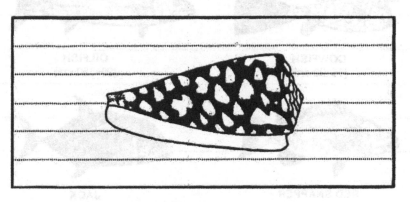

Figure 5-13: Cone shells (Conidae species)

These cone-shaped shells have smooth, colorful mottling and long, narrow openings in the base of the shell. They live under rocks, in crevices or coral reefs, and along rocky shore of protected bays in tropical areas. All have tiny teeth that are similar to hypodermic needles. They can inject an extremely poisonous venom that acts very swiftly, causing acute pain, swelling, paralysis, blindness, and possible death within hours. Avoid handling all cone shells.

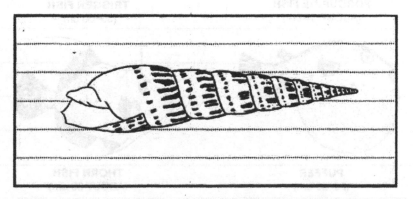

Figure 5-14: Terebra shells or Auger shells (Terebridae species)

These shells are found in both temperate and tropical waters. They are similar to cone shells but much thinner and longer. They poison in the same way as cone shells, but the venom is slightly less poisonous.

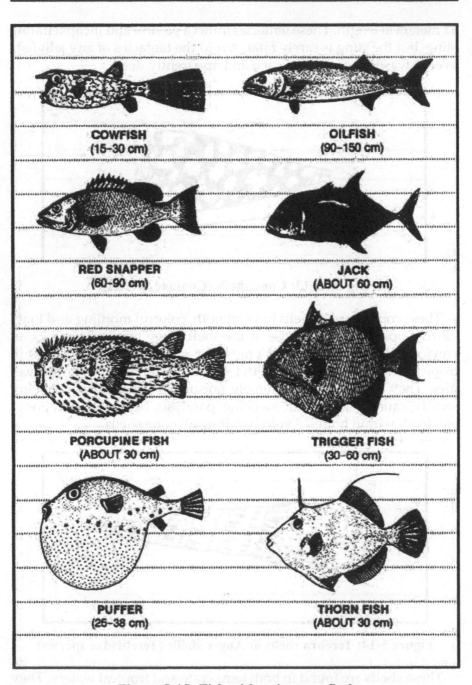

COWFISH
(15–30 cm)

OILFISH
(90–150 cm)

RED SNAPPER
(60–90 cm)

JACK
(ABOUT 60 cm)

PORCUPINE FISH
(ABOUT 30 cm)

TRIGGER FISH
(30–60 cm)

PUFFER
(25–38 cm)

THORN FISH
(ABOUT 30 cm)

Figure 5-15: Fish with poisonous flesh

FISH WITH TOXIC FLESH

There are no simple rules to tell edible fish from those with poisonous flesh. The most common toxic fish are shown in Figure 5-15. All of these fish contain various types of poisonous substances or toxins in their flesh and are dangerous to eat. They have the following common characteristics:

- Most live in shallow water around reefs or lagoons.
- Many have boxy or round bodies with hard shell-like skins covered with bony plates or spines. They have small parrot-like mouths, small gills, and small or absent belly fins. Their names suggest their shape.

Blowfish or puffer (*Tetraodontidae* species) are more tolerant of coldwater. You find them along tropical and temperate coasts worldwide, even in some of the rivers of Southeast Asia and Africa. Stout-bodied and round, many of these fish have short spines and can inflate themselves into a ball when alarmed or agitated. Their blood, liver, and gonads are so toxic that as little as 28 milligrams (1 ounce) can be fatal. These fish vary in color and size, growing up to 75 centimeters in length.

The triggerfish (*Balistidae* species) occur in great variety, mostly in tropical seas. They are deep-bodied and compressed, resembling a seagoing pancake up to 60 centimeters in length, with large and sharp dorsal spines. Avoid them all, as many have poisonous flesh.

Although most people avoid them because of their ferocity, they occasionally eat barracuda (*Sphyraena* barracuda). These predators of mostly tropical seas can reach almost 1.5 meters in length and have attacked humans without provocation. They occasionally carry the poison ciguatera in their flesh, making them deadly if consumed.

In addition to the above fish and their characteristics, red snapper fish may carry ciguatera, a toxin that accumulates in the systems of fish that feed on tropical marine reefs.

Without specific local information, take the following precautions:

- Be very careful with fish taken from normally shallow lagoons with sandy or broken coral bottoms. Reef-feeding species predominate and some may be poisonous.
- Avoid poisonous fish on the leeward side of an island. This area of shallow water consists of patches of living corals

mixed with open spaces and may extend seaward for some distance. Many different types of fish inhabit these shallow waters, some of which are poisonous.

- Do not eat fish caught in any area where the water is unnaturally discolored. This may be indicative of plankton that cause various types of toxicity in plankton-feeding fish.
- Try fishing on the windward side or in deep passages leading from the open sea to the lagoon, but be careful of currents and waves. Live coral reefs drop off sharply into deep water and form a dividing line between the suspected fish of the shallows and the desirable deepwater species. Deepwater fish are usually not poisonous. You can catch the various toxic fish even in deep water. Discard all suspected reef fish, whether caught on the ocean or the reef side.

CHAPTER 6

SURVIVAL USE OF PLANTS

After having solved the problems of finding water, shelter, and animal food, you will have to consider the use of plants you can eat. In a survival situation you should always be on the lookout for familiar wild foods and live off the land whenever possible.

You must not count on being able to go for days without food as some sources would suggest. Even in the most static survival situation, maintaining health through a complete and nutritious diet is essential to maintaining strength and peace of mind.

Nature can provide you with food that will let you survive any ordeal, if you don't eat the wrong plant. You must therefore learn as much as possible beforehand about the flora of the region where you will be operating. Plants can provide you with medicines in a survival situation. Plants can supply you with weapons and raw materials to construct shelters and build fires. Plants can even provide you with chemicals for poisoning fish, preserving animal hides, and for camouflaging yourself and your equipment.

Note: You will find illustrations of the plants described at the end of this chapter and the end of Chapter 7.

EDIBILITY OF PLANTS

Plants are valuable sources of food because they are widely available, easily procured, and, in the proper combinations, can meet all your nutritional needs.

WARNING

The critical factor in using plants for food is to avoid accidental poisoning. Eat only those plants you can positively identify and you know are safe to eat. See Chapter 7, Poisonous Plants, for more information.

Absolutely identify plants before using them as food. Poison hemlock has killed people who mistook it for its relatives, wild carrots and wild parsnips.

At times you may find yourself in a situation for which you could not plan. In this instance you may not have had the chance to learn the plant life of the region in which you must survive. In this case you can use the Universal Edibility Test to determine which plants you can eat and those to avoid.

It is important to be able to recognize both cultivated and wild edible plants in a survival situation. Most of the information in this chapter is directed towards identifying wild plants because information relating to cultivated plants is more readily available.

Remember the following when collecting wild plants for food:

- Plants growing near homes and occupied buildings or along roadsides may have been sprayed with pesticides. Wash them thoroughly. In more highly developed countries with many automobiles, avoid roadside plants, if possible, due to contamination from exhaust emissions.
- Plants growing in contaminated water or in water containing *Giardia lamblia* and other parasites are contaminated themselves. Boil or disinfect them.
- Some plants develop extremely dangerous fungal toxins. To lessen the chance of accidental poisoning, do not eat any fruit that is starting to spoil or showing signs of mildew or fungus.
- Plants of the same species may differ in their toxic or subtoxic compounds content because of genetic or environmental factors. One example of this is the foliage of the common chokecherry. Some chokecherry plants have high concentrations of deadly cyanide compounds while others have low concentrations or none. Horses have died from eating wilted wild cherry leaves. Avoid any weed, leaves, or seeds with an almondlike scent, a characteristic of the cyanide compounds.
- Some people are more susceptible to gastric distress (from plants) than others. If you are sensitive in this way, avoid unknown wild plants. If you are extremely sensitive to poison ivy, avoid products from this family, including any parts from sumacs, mangoes, and cashews.
- Some edible wild plants, such as acorns and water lily rhizomes, are bitter. These bitter substances, usually tannin

compounds, make them unpalatable. Boiling them in several changes of water will usually remove these bitter properties.

- Many valuable wild plants have high concentrations of oxalate compounds, also known as oxalic acid. Oxalates produce a sharp burning sensation in your mouth and throat and damage the kidneys. Baking, roasting, or drying usually destroys these oxalate crystals. The corm (bulb) of the jack-in-the-pulpit is known as the "Indian turnip," but you can eat it only after removing these crystals by slow baking or by drying.

> ## WARNING
> Do not eat mushrooms in a survival situation! The only way to tell if a mushroom is edible is by positive identification. There is no room for experimentation. Symptoms of the most dangerous mushrooms affecting the central nervous system may show up after several days have passed, when it is too late to reverse their effects.

Plant Identification

You identify plants, other than by memorizing particular varieties through familiarity, by using such factors as leaf shape and margin, leaf arrangements, and root structure.

The basic leaf margins (Figure 6-1) are toothed, lobed, and toothless or smooth.

These leaves may be lance-shaped, elliptical, egg-shaped, oblong, wedge-shaped, triangular, long-pointed, or top-shaped (Figure 6-2).

The basic types of leaf arrangements (Figure 6-3) are opposite, alternate, compound, simple, and basal rosette.

The basic types of root structures (Figure 6-4) are the bulb, clove, taproot, tuber, rhizome, corm, and crown. Bulbs are familiar to us as onions and, when sliced in half, will show concentric rings. Cloves are those bulblike structures that remind us of garlic and will separate into small pieces when broken apart. This characteristic separates wild onions from wild garlic. Taproots resemble carrots and may be single-rooted or branched, but usually only one plant stalk arises from each root. Tubers are like potatoes and daylilies and you will find these structures either on strings or in clusters underneath the parent plants. Rhizomes are large creeping rootstock or underground

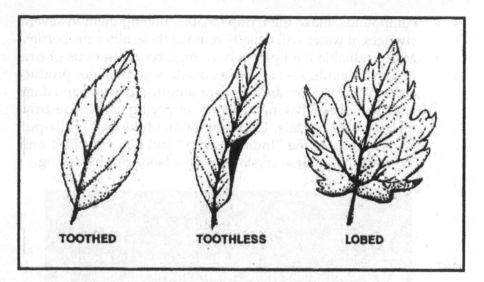

Figure 6-1: Leaf margins

stems and many plants arise from the "eyes" of these roots. Corms are similar to bulbs but are solid when cut rather than possessing rings. A crown is the type of root structure found on plants such as asparagus and looks much like a mophead under the soil's surface.

Learn as much as possible about plants you intend to use for food and their unique characteristics. Some plants have both edible and poisonous parts. Many are edible only at certain times of the year. Others may have poisonous relatives that look very similar to the ones you can eat or use for medicine.

Universal Edibility Test
There are many plants throughout the world. Tasting or swallowing even a small portion of some can cause severe discomfort, extreme internal disorders, and even death. Therefore, if you have the slightest doubt about a plant's edibility, apply the Universal Edibility Test (Table 6-1) before eating any portion of it.

Before testing a plant for edibility, make sure there are enough plants to make the testing worth your time and effort. Each part of a plant (roots, leaves, flowers, and so on) requires more than 24 hours to test. Do not waste time testing a plant that is not relatively abundant in the area.

Remember, eating large portions of plant food on an empty stomach may cause diarrhea, nausea, or cramps. Two good examples of

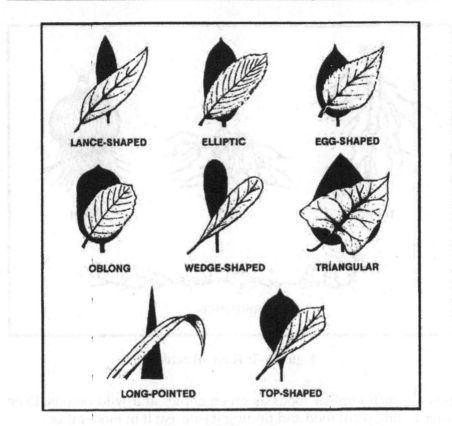

Figure 6-2: Leaf shapes

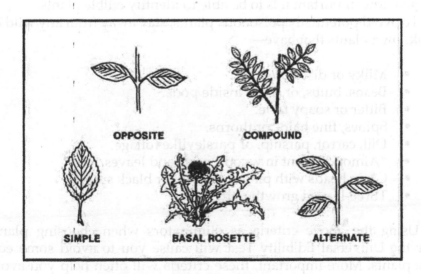

Figure 6-3: Leaf arrangements

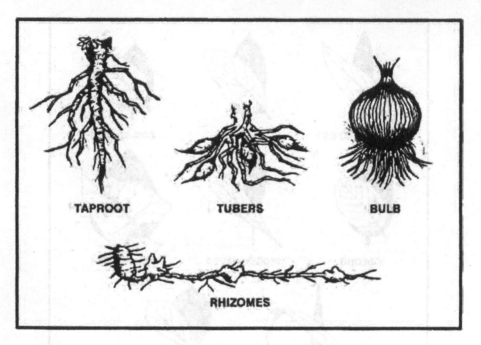

Figure 6-4: Root structures

this are such familiar foods as green apples and wild onions. Even after testing plant food and finding it safe, eat it in moderation.

You can see from the steps and time involved in testing for edibility just how important it is to be able to identify edible plants.

To avoid potentially poisonous plants, stay away from any wild or unknown plants that have—

- Milky or discolored sap.
- Beans, bulbs, or seeds inside pods.
- Bitter or soapy taste.
- Spines, fine hairs, or thorns.
- Dill, carrot, parsnip, or parsleylike foliage.
- "Almond" scent in woody parts and leaves.
- Grain heads with pink, purplish, or black spurs.
- Three-leaved growth pattern.

Using the above criteria as eliminators when choosing plants for the Universal Edibility Test will cause you to avoid some edible plants. More important, these criteria will often help you avoid plants that are potentially toxic to eat or touch.

Table 6-1: Universal Edibility Test

1	Test only one part of a potential food plant at a time.
2	Separate the plant into its basic components—leaves, stems, roots, buds, and flowers.
3	Smell the food for strong or acid odors. Remember, smell alone does not indicate a plant is edible or inedible.
4	Do not eat for 8 hours before starting the test.
5	During the 8 hours you abstain from eating, test for contact poisoning by placing a piece of the plant part you are testing on the inside of your elbow or wrist. Usually 15 minutes is enough time to allow for a reaction.
6	During the test period, take nothing by mouth except purified water and the plant part you are testing.
7	Select a small portion of a single part and prepare it the way you plan to eat it.
8	Before placing the prepared plant part in your mouth, touch a small portion (a pinch) to the outer surface of your lip to test for burning or itching.
9	If after 3 minutes there is no reaction on your lip, place the plant part on your tongue, holding it there for 15 minutes.
10	If there is no reaction, thoroughly chew a pinch and hold it in your mouth for 15 minutes. Do not swallow.
11	If no burning, itching, numbing, stinging, or other irritation occurs during the 15 minutes, swallow the food.
12	Wait 8 hours. If any ill effects occur during this period, induce vomiting and drink a lot of water.
13	If no ill effects occur, eat 0.25 cup of the same plant part prepared the same way. Wait another 8 hours. If no ill effects occur, the plant part as prepared is safe for eating.

CAUTION

Test all parts of the plant for edibility, as some plants have both edible and inedible parts. Do not assume that a part that proved edible when cooked is also edible when raw. Test the part raw to ensure edibility before eating raw. The same part or plant may produce varying reactions in different individuals.

An entire encyclopedia of edible wild plants could be written, but space limits the number of plants presented here. Learn as much as possible about the plant life of the areas where you train regularly and where you expect to be traveling or working. Listed below and on the following pages are some of the most common edible and medicinal plants. Detailed descriptions and photographs of these and other common plants are at the end of this chapter.

Table 6-2: TEMPERATE FOOD ZONE PLANTS

TEMPERATE ZONE FOOD PLANTS
• Amaranth (*Amaranthus retrof/exus* and other species)
• Arrowroot (*Sagittaria species*)
• Asparagus (*Asparagus officinalis*)
• Beechnut (*Fagus* species)
• Blackberries (*Rubus* species)
• Blueberries (*Vaccinium* species)
• Burdock (*Arctium lappa*)
• Cattail (*Typha* species)
• Chestnut (*Castanea* species)
• Chicory (*Cichorium intybus*)
• Chufa (*Cyperus esculentus*)
• Dandelion (*Taraxacum officinale*)
• Daylily (*Hemerocallis fulva*)
• Nettle (*Utica* species)
• Oaks (*Quercus* species)
• Persimmon (*Diospyros virginiana*)
• Plantain (*Plantago* species)
• Pokeweed (*Phytolacca americana*)
• Prickly pear cactus (*Opuntia species*)
• Purslane (*Portulaca oleracea*)
• Sassafras (*Sassafras albidum*)
• Sheep sorrel (*Rumex acetosella*)
• Strawberries (*Fragaria* species)
• Thistle (*Cirsium* species)
• Water lily and lotus (*Nuphar, Nelumbo*, and other species)
• Wild onion and garlic (*Allium* species)
• Wild rose (*Rosa* species)
• Wood sorrel (*Oxalis* species)

Table 6-3: TROPICAL ZONE FOOD PLANTS

TROPICAL ZONE FOOD PLANTS
• Bamboo *(Bambusa* and other species)
• Bananas (Musa species)
• Breadfruit *(Artocarpus incisa)*
• Cashew nut *(Anacardium occidental)*
• Coconut (Cocos *nucifera)*
• Mango *(Mangifera indica)*
• Palms (various species)
• Papaya *(Carica* species)
• Sugarcane *(Saccharum officinarum)*
• Taro *(Colocasia* species)

Table 6-4: DESERT ZONE FOOD PLANTS

DESERT ZONE FOOD PLANTS
• Acacia *(Acacia farnesiana)*
• Agave *(Agave* species)
• Cactus (various species)
• Date palm *(Phoenix dactylifera)*
• Desert amaranth *(Amaranths palmeri)*

Seaweeds

One plant you should never overlook is seaweed. It is a form of marine algae found on or near ocean shores. There are also some edible freshwater varieties. Seaweed is a valuable source of iodine, other minerals, and vitamin C. Large quantities of seaweed in an unaccustomed stomach can produce a severe laxative effect.

When gathering seaweeds for food, find living plants attached to rocks or floating free. Seaweed washed onshore any length of time may be spoiled or decayed. You can dry freshly harvested seaweeds for later use.

Its preparation for eating depends on the type of seaweed. You can dry thin and tender varieties in the sun or over a fire until crisp. Crush and add these to soups or broths. Boil thick, leathery seaweeds for a short time to soften them. Eat them as a vegetable or with other foods. You can eat some varieties raw after testing for edibility.

Table 6-5: SEAWEEDS

SEAWEEDS
• Dulse *(Rhodymenia palmata)*
• Green seaweed *(Ulva lactuca)*
• Irish moss *(Chondrus crispus)*
• Kelp *(Alaria esculenta)*
• Laver *(Porphyra* species)
• Mojaban *(Sargassum fulvellum)*
• Sugar wrack *(Laminaria saccharin)*

Preparation of Plant Food

Although some plants or plant parts are edible raw, you must cook others to be edible or palatable. Edible means that a plant or food will provide you with necessary nutrients, while palatable means that it actually is pleasing to eat. Many wild plants are edible but barely palatable. It is a good idea to learn to identify, prepare, and eat wild foods.

Methods used to improve the taste of plant food include soaking, boiling, cooking, or leaching. Leaching is done by crushing the food (for example, acorns), placing it in a strainer, and pouring boiling water through it or immersing it in running water.

Boil leaves, stems, and buds until tender, changing the water, if necessary, to remove any bitterness.

Boil, bake, or roast tubers and roots. Drying helps to remove caustic oxalates from some roots like those in the Arum family.

Leach acorns in water, if necessary, to remove the bitterness. Some nuts, such as chestnuts, are good raw, but taste better roasted.

You can eat many grains and seeds raw until they mature. When hard or dry, you may have to boil or grind them into meal or flour.

The sap from many trees, such as maples, birches, walnuts, and sycamores, contains sugar. You may boil these saps down to a syrup for sweetening. It takes about 35 liters of maple sap to make one liter of maple syrup!

PLANTS FOR MEDICINE

In a survival situation you will have to use what is available. In using plants and other natural remedies, positive identification of the plants involved is as critical as in using them for food. Proper use of these plants is equally important.

Terms and Definitions

- The following terms, and their definitions, are associated with medicinal plant use: Poultice. The name given to crushed leaves or other plant parts, possibly heated, that you apply to a wound or sore either directly or wrapped in cloth or paper.
- Infusion or tisane or tea. The preparation of medicinal herbs for internal or external application. You place a small quantity of an herb in a container, pour hot water over it, and let it steep (covered or uncovered) before use.
- Decoction. The extract of a boiled down or simmered herb leaf or root. You add herb leaf or root to water. You bring them to a sustained boil or simmer to draw their chemicals into the water. The average ratio is about 28 to 56 grams (1 to 2 ounces) of herb to 0.5 liter of water.
- Expressed juice. Liquids or saps squeezed from plant material and either applied to the wound or made into another medicine.

Many natural remedies work slower than the medicines you know. Therefore, start with smaller doses and allow more time for them to take effect. Naturally, some will act more rapidly than others.

Specific Remedies
The following remedies are for use only in a survival situation, not for routine use:

- Diarrhea. Drink tea made from the roots of blackberries and their relatives to stop diarrhea. White oak bark and other barks containing tannin are also effective. However, use them with caution when nothing else is available because of possible negative effects on the kidneys. You can also stop diarrhea by eating white clay or campfire ashes. Tea made from cowberry or cranberry or hazel leaves works too.
- Antihemorrhagics. Make medications to stop bleeding from a poultice of the puffball mushroom, from plantain leaves, or most effectively from the leaves of the common yarrow or woundwort (*Achilles millefolium*).
- Antiseptics. Use to cleanse wounds, sores, or rashes. You can make them from the expressed juice from wild onion

or garlic, or expressed juice from chickweed leaves or the crushed leaves of dock. You can also make antiseptics from a decoction of burdock root, mallow leaves or roots, or white oak bark. All these medications are for external use only.

- Fevers. Treat a fever with a tea made from willow bark, an infusion of elder flowers or fruit, linden flower tea, or elm bark decoction.
- Colds and sore throats. Treat these illnesses with a decoction made from either plantain leaves or willow bark. You can also use a tea made from burdock roots, mallow or mullein flowers or roots, or mint leaves.
- Aches, pains, and sprains. Treat with externally applied poultices of dock, plantain, chickweed, willow bark, garlic, or sorrel. You can also use salves made by mixing the expressed juices of these plants in animal fat or vegetable oils.
- Itching. Relieve the itch from insect bites, sunburn, or plant poisoning rashes by applying a poultice of jewelweed (*Impatiens biflora*) or witch hazel leaves (*Hamamelis virginiana*).The jewelweed juice will help when applied to poison ivy rashes or insect stings. It works on sunburn as well as aloe vera.
- Sedatives. Get help in falling asleep by brewing a tea made from mint leaves or passionflower leaves.
- Hemorrhoids. Treat them with external washes from elm bark or oak bark tea, from the expressed juice of plantain leaves, or from a Solomon's seal root decoction.
- Constipation. Relieve constipation by drinking decoctions from dandelion leaves, rose hips, or walnut bark. Eating raw daylily flowers will also help.
- Worms or intestinal parasites. Using moderation, treat with tea made from tansy (*Tanacetum vulgare*) or from wild carrot leaves.
- Gas and cramps. Use a tea made from carrot seeds as an antiflatulent; use tea made from mint leaves to settle the stomach.
- Antifungal washes. Make a decoction of walnut leaves or oak bark or acorns to treat ringworm and athlete's foot. Apply frequently to the site, alternating with exposure to direct sunlight.

MISCELLANEOUS USE OF PLANTS

- Make dyes from various plants to color clothing or to camouflage your skin. Usually, you will have to boil the plants to get the best results. Onion skins produce yellow, walnut hulls produce brown, and pokeberries provide a purple dye.

- Make fibers and cordage from plant fibers. Most commonly used are the stems from nettles and milkweeds, yucca plants, and the inner bark of trees like the linden.
- Make fish poison by immersing walnut hulls in a small area of quiet water. This poison makes it impossible for the fish to breathe but doesn't adversely affect their edibility.
- Make tinder for starting fires from cattail fluff, cedar bark, lighter knot wood from pine trees, or hardened sap from resinous wood trees.
- Make insulation by fluffing up female cattail heads or milkweed down.
- Make insect repellents by applying the expressed juice of wild garlic or onion to the skin, by placing sassafras leaves in your shelter, or by burning or smudging cattail seed hair fibers.

Plants can be your ally as long as you use them cautiously. The key to the safe use of plants is positive identification whether you use them as food or medicine or in constructing shelters or equipment.

EDIBLE AND MEDICINAL PLANTS

Abal

Calligonum comosum

Description: The abal is one of the few shrubby plants that exists in the shady deserts. This plant grows to about 1.2 meters, and its branches look like wisps from a broom. The stiff, green branches produce an abundance of flowers in the early spring months (March, April).

Habitat and Distribution: This plant is found in desert scrub and waste in any climatic zone. It inhabits much of the North African desert. It may also be found on the desert sands of the Middle East and as far eastward as the Rajputana desert of westen India.

Edible Parts: This plant's general appearance would not indicate its usefulness to the survivor, but while this plant is flowering in the spring, its fresh flowers can be eaten. This plant is common in the areas where it is found. An analysis of the food value of this plant has shown it to be high in sugar and nitrogenous components.

Acacia
Acacia farnesiana

Description: Acacia is a spreading, usually short tree with spines and alternate compound leaves. Its individual leaflets are small. Its flowers are ball-shaped, bright yellow, and very fragrant. Its bark is a whitish-gray color. Its fruits are dark brown and podlike.

Habitat and Distribution: Acacia grows in open, sunny areas. It is found throughout all tropical regions.

Note: There are about 500 species of acacia. These plants are especially prevalent in Africa, southern Asia, and Australia, but many species are found in the warmer and drier parts of America.

Edible Parts: Its young leaves, flowers, and pods are edible raw or cooked.

Agave
Agave species

Description: These plants have large clusters of thick, fleshy leaves borne close to the ground and surrounding a central stalk. The plants flower only once, then die. They produce a massive flower stalk.

Habitat and Distribution: Agaves prefer dry, open areas. They are found throughout Central America, the Caribbean, and parts of the western deserts of the United States and Mexico.

Edible Parts: Its flowers and flower buds are edible. Boil them before eating.

⚠ CAUTION
The juice of some species causes dermatitis in some individuals.

Other Uses: Cut the huge flower stalk and collect the juice for drinking. Some species have very fibrous leaves. Pound the leaves and remove the fibers for weaving and making ropes. Most species have thick, sharp needles at the tips of the leaves. Use them for sewing or making hacks. The sap of some species contains a chemical that makes the sap suitable for use as a soap.

Almond
Prunus amygdalus

Description: The almond tree, which sometimes grows to 12.2 meters, looks like a peach tree. The fresh almond fruit resembles a gnarled, unripe peach and grows in clusters. The stone (the almond itself) is covered with a thick, dry, woolly skin.

Habitat and Distribution: Almonds are found in the scrub and thorn forests of the tropics, the evergreen scrub forests of temperate areas, and in desert scrub and waste in all climatic zones. The almond tree is also found in the semidesert areas of the Old World in southern Europe, the eastern Mediterranean, Iran, the Middle East, China, Madeira, the Azores, and the Canary Islands.

Edible Parts: The mature almond fruit splits open lengthwise down the side, exposing the ripe almond nut. You can easily get the dry kernel by simply cracking open the stone. Almond meats are rich in food value, like all nuts. Gather them in large quantities and shell them for further use as survival food. You could live solely on almonds for rather long periods. When you boil them, the kernel's outer covering comes off and only the white meat remains.

Amaranth
Amaranthus species

Description: These plants, which grow 90 centimeters to 150 centimeters tall, are abundant weeds in many parts of the world. All amaranth have alternate simple leaves. They may have some red color present on the stems. They bear minute, greenish flowers in dense clusters at the top of the plants. Their seeds may be brown or black in weedy species and light-colored in domestic species.

Habitat and Distribution: Look for amaranth along roadsides, in disturbed waste areas, or as weeds in crops throughout the world. Some amaranth species have been grown as a grain crop and a garden vegetable in various parts of the world, especially in South America.

Edible Parts: All parts are edible, but some may have sharp spines you should remove before eating. The young plants or the growing tips of alder plants are an excellent vegetable. Simply boil the young plants or eat them raw. Their seeds are very nutritious. Shake the tops of alder plants to get the seeds. Eat the seeds raw, boiled, ground into flour, or popped like popcorn.

Arctic willow
Salix arctica

Description: The arctic willow is a shrub that never exceeds more than 60 centimeters in height and grows in clumps that form dense mats on the tundra.

Habitat and Distribution: The arctic willow is common on tundras in North America. Europe, and Asia. You can also find it in some mountainous areas in temperate regions.

Edible Parts: You can collect the succulent, tender young shoots of the arctic willow in early spring. Strip off the outer bark of the new shoots and eat the inner portion raw. You can also peel and eat raw the young underground shoots of any of the various kinds of arctic

willow. Young willow leaves are one of the richest sources of vitamin C, containing 7 to 10 times more than an orange.

Arrowroot
Maranta and *Sagittaria* species

Description: The arrowroot is an aquatic plant with arrow-shaped leaves and potatolike tubers in the mud.

Habitat and Distribution: Arrowroot is found worldwide in temperate zones and the tropics. It is found in moist to wet habitats.

Edible Parts: The rootstock is a rich source of high quality starch. Boil the rootstock and eat it as a vegetable.

Asparagus
Asparagus officinalis

Description: The spring growth of this plant resembles a cluster of green fingers. The mature plant has fernlike, wispy foliage and red berries. Its flowers are small and greenish in color. Several species have sharp, thornlike structures.

Habitat and Distribution: Asparagus is found worldwide in temperate areas. Look for it in fields, old homesites, and fencerows.

Edible Parts: Eat the young stems before leaves form. Steam or boil them for 10 to 15 minutes before eating. Raw asparagus may cause nausea or diarrhea. The fleshy roots are a good source of starch.

Bael fruit
Aegle marmelos

Description: This is a tree that grows from 2.4 to 4.6 meters tall, with a dense spiny growth. The fruit is 5 to 10 centimeters in diameter, gray or yellowish, and full of seeds.

Habitat and Distribution: Bael fruit is found in rain forests and semi-evergreen seasonal forests of the tropics. It grows wild in India and Burma.

Edible Parts: The fruit, which ripens in December, is at its best when just turning ripe. The juice of the ripe fruit, diluted with water and mixed with a small amount of tamarind and sugar or honey, is sour but refreshing. Like other citrus fruits, it is rich in vitamin C.

Bamboo

Various species including *Bambusa, Dendrocalamus, Phyllostachys*

Description: Bamboos are woody grasses that grow up to 15 meters tall. The leaves are grasslike and the stems are the familiar bamboo used in furniture and fishing poles.

Habitat and Distribution: Look for bamboo in warm, moist regions in open or jungle country, in lowland, or on mountains. Bamboos are native to the Far East (Temperate and Tropical zones) but have bean widely planted around the world.

Edible Parts: The young shoots of almost all species are edible raw or cooked. Raw shoots have a slightly bitter taste that is removed by boiling. To prepare, remove the tough protective sheath that is coated with tawny or red hairs. The seed grain of the flowering bamboo is also edible. Boil the seeds like rice or pulverize them, mix with water, and make into cakes.

Other Uses: Use the mature bamboo to build structures or to make containers, ladles, spoons, and various other cooking utensils. Also use bamboo to make tools and weapons. You can make a strong bow by splitting the bamboo and putting several pieces together.

⚠ CAUTION
Green bamboo may explode in a fire. Green bamboo has an internal membrane you must remove before using it as a food or water container.

Banana and plantain
Musa species

Description: These are treelike plants with several large leaves at the top. Their flowers are borne in dense hanging clusters.

Habitat and Distribution: Look for bananas and plantains in open fields or margins of forests where they are grown as a crop. They grow in the humid tropics.

Edible Parts: Their fruits are edible raw or cooked. They may be boiled or baked. You can boil their flowers and eat them like a vegetable. You can cook and eat the rootstocks and leaf sheaths of many species. The center or "heart" or the plant is edible year-round, cooked or raw.

Other Uses: You can use the layers of the lower third of the plants to cover coals to roast food. You can also use their stumps to get water (see Chapter 6). You can use their leaves to wrap other foods for cooking or storage.

Baobab

Adansonia digitata

Description: The baobab tree may grow as high as 18 meters and may have a trunk 9 meters in diameter. The tree has short, stubby branches and a gray, thick bark. Its leaves are compound and their segments are arranged like the palm of a hand. Its flowers, which are white and several centimeters across, hang from the higher branches. Its fruit is shaped like a football, measures up to 45 centimeters long, and is covered with short dense hair.

Habitat and Distribution: These trees grow in savannas. They are found in Africa, in parts of Australia, and on the island of Madagascar.

Edible Parts: You can use the young leaves as a soup vegetable. The tender root of the young baobab tree is edible. The pulp and seeds of the fruit are also edible. Use one handful of pulp to about one cup of water for a refreshing drink. To obtain flour, roast the seeds, then grind them.

Other Uses: Drinking a mixture of pulp and water will help cure diarrhea. Often the hollow trunks are good sources of fresh water. The bark can be cut into strips and pounded to obtain a strong fiber for making rope.

Batoko plum
Flacourtia inermis

Description: This shrub or small tree has dark green, alternate, simple leaves. Its fruits are bright red and contain six or more seeds.

Habitat and Distribution: This plant is a native of the Philippines but is widely cultivated for its fruit in other areas. It can be found in clearings and at the edges of the tropical rain forests of Africa and Asia.

Edible Parts: Eat the fruit raw or cooked.

Bearberry or kinnikinnick
Arctostaphylos uvaursi

Description: This plant is a common evergreen shrub with reddish, scaly bark and thick, leathery leaves 4 centimeters long and 1 centimeter wide. It has white flowers and bright red fruits.

Habitat and Distribution: This plant is found in arctic, subarctic, and temperate regions, most often in sandy or rocky soil.

Edible Parts: Its berries are edible raw or cooked. You can make a refreshing tea from its young leaves.

Beech
Fagus species

Description: Beech trees are large (9 to 24 meters), symmetrical forest trees that have smooth, light-gray bark and dark green foliage. The character of its bark, plus its clusters of prickly seedpods, clearly distinguish the beech tree in the field.

Habitat and Distribution: This tree is found in the Temperate Zone. It grows wild in the eastern United States, Europe, Asia, and North Africa. It is found in moist areas, mainly in the forests. This tree is common throughout southeastern Europe and across temperate Asia. Beech relatives are also found in Chile, New Guinea, and New Zealand.

Edible Parts: The mature beechnuts readily fall out of the husklike seedpods. You can eat these dark brown triangular nuts by breaking the thin shell with your fingernail and removing the white, sweet kernel inside. Beechnuts are one of the most delicious of all wild nuts. They are a most useful survival food because of the kernel's high oil content. You can also use the beechnuts as a coffee substitute. Roast them so that the kernel becomes golden brown and quite hard. Then pulverize the kernel and, after boiling or steeping in hot water, you have a passable coffee substitute.

Bignay
Antidesma bunius

Description: Bignay is a shrub or small tree, 3 to 12 meters tall, with shiny, pointed leaves about 15 centimeters long. Its flowers are small,

clustered, and green. It has fleshy, dark red or black fruit and a single seed. The fruit is about 1 centimeter in diameter.

Habitat and Distribution: This plant is found in rain forests and semi-evergreen seasonal forests in the tropics. It is found in open places and in secondary forests. It grows wild from the Himalayas to Ceylon and eastward through Indonesia to northern Australia. However, it may be found anywhere in the tropics in cultivated forms.

Edible Parts: The fruit is edible raw. Do not eat any other parts of the tree. In Africa, the roots are toxic. Other parts of the plant may be poisonous.

> ⚠ CAUTION
> Eaten in large quantities, the fruit may have a laxative effect.

Blackberry, raspberry, and dewberry
Rubus species

Description: These plants have prickly stems (canes) that grow upward, arching back toward the ground. They have alternate, usually compound leaves. Their fruits may be red, black, yellow, or orange.

Habitat and Distribution: These plants grow in open, sunny areas at the margin of woods, lakes, streams, and roads throughout temperate regions. There is also an arctic raspberry.

Edible Parts: The fruits and peeled young shoots are edible. Flavor varies greatly.

Other Uses: Use the leaves to make tea. To treat diarrhea, drink a tea made by brewing the dried root bark of the blackberry bush.

Blueberry and huckleberry
Vaccinium and *Gaylussacia* species

Description: These shrubs vary in size from 30 centimeters to 3.7 meters tall. All have alternate, simple leaves. Their fruits may be dark blue, black, or red and have many small seeds.

Habitat and Distribution: These plants prefer open, sunny areas. They are found throughout much of the north temperate regions and at higher elevations in Central America.

Edible Parts: Their fruits are edible raw.

Breadfruit
Artocarpus incisa

Description: This tree may grow up to 9 meters tall. It has dark green, deeply divided leaves that are 75 centimeters long and 30 centimeters wide. Its fruits are large, green, ball-like structures up to 30 centimeters across when mature.

Habitat and Distribution: Look for this tree at the margins of forests and homesites in the humid tropics. It is native to the South Pacific region but has been widely planted in the West Indies and parts of Polynesia.

Edible Parts: The fruit pulp is edible raw. The fruit can be sliced, dried, and ground into flour for later use. The seeds are edible cooked.

Other Uses: The thick sap can serve as glue and caulking material. You can also use it as birdlime (to entrap small birds by smearing the sap on twigs where they usually perch).

Burdock
Arctium lappa

Description: This plant has wavy-edged, arrow-shaped leaves and flower heads in burrlike clusters. It grows up to 2 meters tall, with purple or pink flowers and a large, fleshy root.

Habitat and Distribution: Burdock is found worldwide in the North Temperate Zone. Look for it in open waste areas during the spring and summer.

Edible Parts: Peel the tender leaf stalks and eat them raw or cook them like greens. The roots are also edible boiled or baked.

⚠ CAUTION
Do not confuse burdock with rhubarb, which has poisonous leaves.

Other Uses: A liquid made from the roots will help to produce sweating and increase urination. Dry the root, simmer it in water, strain the liquid, and then drink the strained liquid. Use the fiber from the dried stalk to weave cordage.

Burl Palm
Corypha elata

Description: This tree may reach 18 meters in height. It has large, fan-shaped leaves up to 3 meters long and split into about 100 narrow segments. It bears flowers in huge clusters at the top of the tree. The tree dies after flowering.

Habitat and Distribution: This tree grows in coastal areas of the East Indies.

Edible Parts: The trunk contains starch that is edible raw. The very tip of the trunk is also edible raw or cooked. You can get large quantities of liquid by bruising the flowering stalk. The kernels of the nuts are edible.

⚠ CAUTION

The seed covering may cause dermatitis in some individuals.

Other Uses: You can use the leaves as weaving material.

Canna lily
Canna indica

Description: The canna lily is a coarse perennial herb, 90 centimeters to 3 meters tall. The plant grows from a large, thick, underground rootstock that is edible. Its large leaves resemble those of the banana plant but are not so large. The flowers of wild canna lily are usually small, relatively inconspicuous, and brightly colored reds, oranges, or yellows.

Habitat and Distribution: As a wild plant, the canna lily is found in all tropical areas, especially in moist places along streams, springs, ditches, and the margins of woods. It may also be found in wet temperate, mountainous regions. It is easy to recognize because it is commonly cultivated in flower gardens in the United States.

Edible Parts: The large and much branched rootstocks are full of edible starch. The younger parts may be finely chopped and then boiled or pulverized into a meal. Mix in the young shoots of palm cabbage for flavoring.

Carob tree
Ceratonia siliqua

Description: This large tree has a spreading crown. Its leaves are compound and alternate. Its seedpods, also known as Saint John's

bread, are up to 45 centimeters long and are filled with round, hard seeds and a thick pulp.

Habitat and Distribution: This tree is found throughout the Mediterranean, the Middle East, and parts of North Africa.

Edible Parts: The young tender pods are edible raw or boiled. You can pulverize the seeds in mature pods and cook as porridge.

Cashew nut
Anacardium occidentale

Description: The cashew is a spreading evergreen tree growing to a height of 12 meters, with leaves up to 20 centimeters long and 10 centimeters wide. Its flowers are yellowish-pink. Its fruit is very easy to recognize because of its peculiar structure. The fruit is thick and pear-shaped, pulpy and red or yellow when ripe. This fruit bears a hard, green, kidney-shaped nut at its tip. This nut is smooth, shiny, and green or brown according to its maturity.

Habitat and Distribution: The cashew is native to the West Indies and northern South America, but transplantation has spread it to all tropical climates. In the Old World, it has escaped from cultivation and appears to be wild at least in parts of Africa and India.

Edible Parts: The nut encloses one seed. The seed is edible when roasted. The pear-shaped fruit is juicy, sweet-acid, and astringent. It is quite safe and considered delicious by most people who eat it.

> ⚠ CAUTION
> The green hull surrounding the nut contains a resinous irritant poison that will blister the lips and tongue like poison ivy. Heat destroys this poison when roasting the nuts.

Cattail
Typha latifolia

Description: Cattails are grasslike plants with strap-shaped leaves 1 to 5 centimeters wide and growing up to 1.8 meters tall. The male flowers are borne in a dense mass above the female flowers. These last only a short time, leaving the female flowers that develop into the brown cattail. Pollen from the male flowers is often abundant and bright yellow.

Habitat and Distribution: Cattails are found throughout most of the world. Look for them in full sun areas at the margins of lakes, streams, canals, rivers, and brackish water.

Edible Parts: The young tender shoots are edible raw or cooked. The rhizome is often very tough but is a rich source of starch. Pound the rhizome to remove the starch and use as a flour. The pollen is also an exceptional source of starch. When the cattail is immature and still green, you can boil the female portion and eat it like corn on the cob.

Other Uses: The dried leaves are an excellent source of weaving material you can use to make floats and rafts. The cottony seeds make good pillow stuffing and insulation. The fluff makes excellent tinder. Dried cattails are effective insect repellents when burned.

Cereus cactus
Cereus species

Description: These cacti are tall and narrow with angled stems and numerous spines.

Habitat and Distribution: They may be found in true deserts and other dry, open, sunny areas throughout the Caribbean region, Central America, and the western United States.

Edible Parts: The fruits are edible, but some may have a laxative effect.

Other Uses: The pulp of the cactus is a good source of water. Break open the stem and scoop out the pulp.

Chestnut
Castanea sativa

Description: The European chestnut is usually a large tree, up to 18 meters in height.

Habitat and Distribution: In temperate regions, the chestnut is found in both hardwood and coniferous forests. In the tropics, it is found in

semi-evergreen seasonal forests. They are found over all of middle and south Europe and across middle Asia to China and Japan. They are relatively abundant along the edge of meadows and as a forest tree. The European chestnut is one of the most common varieties. Wild chestnuts in Asia belong to the related chestnut species.

Edible Parts: Chestnuts are highly useful as survival food. Ripe nuts are usually picked in autumn, although unripe nuts picked while green may also be used for food. Perhaps the easiest way to prepare them is to roast the ripe nuts in embers. Cooked this way, they are quite tasty, and you can eat large quantities. Another way is to boil the kernels after removing the outer shell. After being boiled until fairly soft, you can mash the nuts like potatoes.

Chicory
Cichorium intybus

Description: This plant grows up to 1.8 meters tall. It has leaves clustered at the base of the stem and some leaves on the stem. The base leaves resemble those of the dandelion.

The flowers are sky blue and stay open only on sunny days. Chicory has a milky juice.

Habitat and Distribution: Look for chicory in old fields, waste areas, weedy lots, and along roads. It is a native of Europe and Asia, but is

also found in Europe and across middle Asia to China and Japan. They are relative abundant along the edge of meadows and as a forest weed.

Edible Parts: All parts are edible. Eat the young leaves as a salad or boil to eat as a vegetable. Cook the roots as a vegetable. For use as a coffee substitute, roast the roots until they are dark brown and then pulverize them.

Chufa
Cyperus esculentus

Description: This very common plant has a triangular stem and grasslike leaves. It grows to a height of 20 to 60 centimeters. The mature plant has a soft furlike bloom that extends from a whorl of leaves. Tubers 1 to 2.5 centimeters in diameter grow at the ends of the roots.

Habitat and Distribution: Chufa grows in moist sandy areas throughout the world. It is often an abundant weed in cultivated fields.

Edible Parts: The tubers are edible raw, boiled, or baked. You can also grind them and use them as a coffee substitute.

Coconut
Cocos nucifera

Description: This tree has a single, narrow, tall trunk with a cluster of very large leaves at the top. Each leaf may be over 6 meters long with over 100 pairs of leaflets.

Habitat and Distribution: Coconut palms are found throughout the tropics. They are most abundant near coastal regions.

Edible Parts: The nut is a valuable source of food. The milk of the young coconut is rich in sugar and vitamins and is an excellent source of liquid. The nut meat is also nutritious but is rich in oil. To preserve the meat, spread it in the sun until it is completely dry.

Other Uses: Use coconut oil to cook and to protect metal objects from corrosion. Also use the oil to treat saltwater sores, sunburn, and dry skin. Use the oil in improvised torches. Use the tree trunk as building

material and the leaves as thatch. Hollow out the large stump for use as a food container. The coconut husks are good flotation devices and the husk's fibers are used to weave ropes and other items. Use the gauzelike fibers at the leaf bases as strainers or use them to weave a bug net or to make a pad to use on wounds. The husk makes a good abrasive. Dried husk fiber is an excellent tinder. A smoldering husk helps to repel mosquitoes. Smoke caused by dripping coconut oil in a fire also repels mosquitoes. To render coconut oil, put the coconut meat in the sun, heat it over a slow fire, or boil it in a pot of water. Coconuts washed out to sea are a good source of fresh liquid for the sea survivor.

Common jujube
Ziziphus jujuba

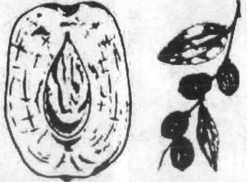

Description: The common jujube is either a deciduous tree growing to a height of 12 meters or a large shrub, depending upon where it grows and how much water is available for growth. Its branches are usually spiny. Its reddish-brown to yellowish-green fruit is oblong to ovoid, 3 centimeters or less in diameter, smooth, and sweet in flavor, but has rather dry pulp around a comparatively large stone. Its flowers are green.

Habitat and Distribution: The jujube is found in forested areas of temperate regions and in desert scrub and waste areas worldwide. It is common in many of the tropical and subtropical areas of the Old World. In Africa, it is found mainly bordering the Mediterranean. In Asia, it is especially common in the drier parts of India and China. The jujube is also found throughout the East Indies. It can be found bordering some desert areas.

Edible Parts: The pulp, crushed in water, makes a refreshing beverage. If time permits, you can dry the ripe fruit in the sun like dates. Its fruits are high in vitamins A and C.

Cranberry
Vaccinium macrocarpon

Description: This plant has tiny leaves arranged alternately. Its stem creeps along the ground. Its fruits are red berries.

Habitat and Distribution: It only grows in open, sunny, wet areas in the colder regions of the Northern Hemisphere.

Edible Parts: The berries are very tart when eaten raw. Cook in a small amount of water and add sugar, if available, to make a jelly.

Other Uses: Cranberries may act as a diuretic. They are useful for treating urinary tract infections.

Crowberry
Empetrum nigrum

Description: This is a dwarf evergreen shrub with short needlelike leaves. It has small, shiny, black berries that remain on the bush throughout the winter.

Habitat and Distribution: Look for this plant in tundra throughout arctic regions of North America and Eurasia.

Edible Parts: The fruits are edible fresh or can be dried for later use.

Cuipo tree
Cavanillesia platanifolia

Description: This is a very dominant and easily detected tree because it extends above the other trees. Its height ranges from 45 to 60 meters. It has leaves only at the top and is bare 11 months out of the year. It has rings on its bark that extend to the top to make it easily

recognizable. Its bark is reddish or gray in color. Its roots are light reddish-brown or yellowish-brown.

Habitat and Distribution: The cuipo tree is located primarily in Central American tropical rain forests in mountainous areas.

Edible Parts: To get water from this tree, cut a piece of the root and clean the dirt and bark off one end, keeping the root horizontal. Put the clean end to your mouth or canteen and raise the other. The water from this tree tastes like potato water.

Other Uses: Use young saplings and the branches' inner bark to make rope.

Dandelion
Taraxacum officinale

Description: Dandelion leaves have a jagged edge, grow close to the ground, and are seldom more than 20 centimeters long. Its flowers are bright yellow. There are several dandelion species.

Habitat and Distribution: Dandelions grow in open, sunny locations throughout the Northern Hemisphere.

Edible Parts: All parts are edible. Eat the leaves raw or cooked. Boil the roots as a vegetable. Roots roasted and ground are a good coffee substitute. Dandelions are high in vitamins A and C and in calcium.

Other Uses: Use the white juice in the flower stems as glue.

Date palm
Phoenix dactylifera

Description: The date palm is a tall, unbranched tree with a crown of huge, compound leaves. Its fruit is yellow when ripe.

Habitat and Distribution: This tree grows in arid semitropical regions. It is native to North Africa and the Middle East but has been planted in the arid semitropics in other parts of the world.

Edible Parts: Its fruit is edible fresh but is very bitter if eaten before it is ripe. You can dry the fruits in the sun and preserve them for a long time.

Other Uses: The trunks provide valuable building material in desert regions where few other treelike plants are found. The leaves are durable and you can use them for thatching and as weaving material. The base of the leaves resembles coarse cloth that you can use for scrubbing and cleaning.

Daylily
Hemerocallis fulva

Description: This plant has unspotted, tawny blossoms that open for 1 day only. It has long, swordlike, green basal leaves. Its root is a mass of swollen and elongated tubers.

Habitat and Distribution: Daylilies are found worldwide in Tropic and Temperate Zones. They are grown as a vegetable in the Orient and as an ornamental plant elsewhere.

Edible Parts: The young green leaves are edible raw or cooked. Tubers are also edible raw or cooked. You can eat its flowers raw, but they taste better cooked. You can also fry the flowers for storage.

> ⚠ **CAUTION**
>
> Eating excessive amounts of raw flowers may cause diarrhea.

Duchesnea or Indian strawberry
Duchesnea indica

Description: The duchesnea is a small plant that has runners and three-parted leaves. Its flowers are yellow and its fruit resembles a strawberry.

Habitat and Distribution: It is native to southern Asia but is a common weed in warmer temperate regions. Look for it in lawns, gardens, and along roads.

Edible Parts: Its fruit is edible. Eat it fresh.

Elderberry
Sambucus canadensis

Description: Elderberry is a many-stemmed shrub with opposite, compound leaves. It grows to a height of 6 meters. Its flowers are fragrant, white, and borne in large flat-topped clusters up to 30 centimeters across. Its berrylike fruits are dark blue or black when ripe.

Habitat and Distribution: This plant is found in open, usually wet areas at the margins of marshes, rivers, ditches, and lakes. It grows throughout much of eastern North America and Canada.

Edible Parts: The flowers and fruits are edible. You can make a drink by soaking the flower heads for 8 hours, discarding the flowers, and drinking the liquid.

> ⚠ CAUTION
> All other parts of the plant are poisonous and danger-ous if eaten.

Fireweed
Epilobium angustifolium

Description: This plant grows up to 1.8 meters tall. It has large, showy, pink flowers and lance-shaped leaves. Its relative, the dwarf fireweed (*Epilobium latifolium*), grows 30 to 60 centimeters tall.

Habitat and Distribution: Tall fireweed is found in open woods, on hillsides, on stream banks, and near seashores in arctic regions. It is especially abundant in burned-over areas. Dwarf fireweed is found along streams, sandbars, and lakeshores and on alpine and arctic slopes.

Edible Parts: The leaves, stems, and flowers are edible in the spring but become tough in summer. You can split open the stems of old plants and eat the pith raw.

Fishtail palm
Caryota urens

Description: Fishtail palms are large trees, at least 18 meters tall. Their leaves are unlike those of any other palm; the leaflets are irregular and toothed on the upper margins. All other palms have either fan-shaped or featherlike leaves. Its massive flowering shoot is borne at the top of the tree and hangs downward.

Habitat and Distribution: The fishtail palm is native to the tropics of India, Assam, and Burma. Several related species also exist in Southeast Asia and the Philippines. These palms are found in open hill country and jungle areas.

Edible Parts: The chief food in this palm is the starch stored in large quantities in its trunk. The juice from the fishtail palm is very

nourishing and you have to drink it shortly after getting it from the palm flower shoot. Boil the juice down to get a rich sugar syrup. Use the same method as for the sugar palm to get the juice. The palm cabbage may be eaten raw or cooked.

Foxtail grass
Setaria species

Description: This weedy grass is readily recognized by the narrow, cylindrical head containing long hairs. Its grains are small, less than 6 millimeters long. The dense heads of grain often droop when ripe.

Habitat and Distribution: Look for foxtail grasses in open, sunny areas, along roads, and at the margins of fields. Some species occur in wet, marshy areas. Species of Setaria are found throughout the United States, Europe, western Asia, and tropical Africa. In some parts of the world, foxtail grasses are grown as a food crop.

Edible Parts: The grains are edible raw but are very hard and sometimes bitter. Boiling removes some of the bitterness and makes them easier to eat.

Goa bean
Psophocarpus tetragonolobus

Description: The goa bean is a climbing plant that may cover small shrubs and trees. Its bean pods are 22 centimeters long, its leaves 15 centimeters long, and its flowers are bright blue. The mature pods are 4-angled, with jagged wings on the pods.

Habitat and Distribution: This plant grows in tropical Africa, Asia, the East Indies, the Philippines, and Taiwan. This member of the bean (legume) family serves to illustrate a kind of edible bean common in

the tropics of the Old World. Wild edible beans of this sort are most frequently found in clearings and around abandoned garden sites. They are more rare in forested areas.

Edible Parts: You can eat the young pods like string beans. The mature seeds are a valuable source of protein after parching or roasting them over hot coals. You can germinate the seeds (as you can many kinds of beans) in damp moss and eat the resultant sprouts. The thickened roots are edible raw. They are slightly sweet, with the firmness of an apple. You can also eat the young leaves as a vegetable, raw or steamed.

Hackberry
Celtis species

Description: Hackberry trees have smooth, gray bark that often has corky warts or ridges. The tree may reach 39 meters in height. Hackberry trees have long-pointed leaves that grow in two rows. This tree bears small, round berries that can be eaten when they are ripe and fall from the tree. The wood of the hackberry is yellowish.

Habitat and Distribution: This plant is widespread in the United States, especially in and near ponds.

Edible Parts: Its berries are edible when they are ripe and fall from the tree.

Hazelnut or wild filbert
Corylus species

Description: Hazelnuts grow on bushes 1.8 to 3.6 meters high. One species in Turkey and another in China are large trees. The nut itself grows in a very bristly husk that conspicuously contracts above the nut into a long neck. The different species vary in this respect as to size and shape.

Habitat and Distribution: Hazelnuts are found over wide areas in the United States, especially the eastern half of the country and along the Pacific coast. These nuts are also found in Europe where they are known as filberts. The hazelnut is common in Asia, especially in

eastern Asia from the Himalayas to China and Japan. The hazelnut usually grows in the dense thickets along stream banks and open places. They are not plants of the dense forest.

Edible Parts: Hazelnuts ripen in the autumn when you can crack them open and eat the kernel. The dried nut is extremely delicious. The nut's high oil content makes it a good survival food. In the unripe stage, you can crack them open and eat the fresh kernel.

Horseradish tree
Moringa pterygosperma

Description: This tree grows from 4.5 to 14 meters tall. Its leaves have a fernlike appearance. Its flowers and long, pendulous fruits grow on the ends of the branches. Its fruit (pod) looks like a giant bean. Its

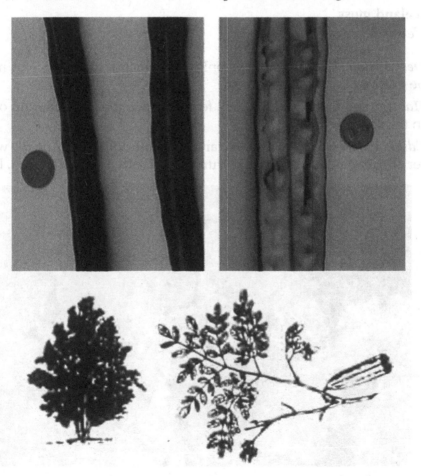

25- to 60-centimeter-long pods are triangular in cross section, with strong ribs. Its roots have a pungent odor.

Habitat and Distribution: This tree is found in the rain forests and semi-evergreen seasonal forests of the tropical regions. It is widespread in India, Southeast Asia, Africa, and Central America. Look for it in abandoned fields and gardens and at the edges of forests.

Edible Parts: The leaves are edible raw or cooked, depending on their hardness. Cut the young seedpods into short lengths and cook them like string beans or fry them. You can get oil for frying by boiling the young fruits of palms and skimming the oil off the surface of the water. You can eat the flowers as part of a salad. You can chew fresh, young seedpods to eat the pulpy and soft seeds. The roots may be ground as a substitute for seasoning similar to horseradish.

Iceland moss
Cetraria islandica

Description: This moss grows only a few inches high. Its color may be gray, white, or even reddish.

Habitat and Distribution: Look for it in open areas. It is found only in the arctic.

Edible Parts: All parts of the Iceland moss are edible. During the winter or dry season, it is dry and crunchy but softens when soaked. Boil

the moss to remove the bitterness. After boiling, eat by itself or add to milk or grains as a thickening agent. Dried plants store well.

Indian potato or Eskimo potato
Claytonia species

Description: All Claytonia species are somewhat fleshy plants only a few centimeters tall, with showy flowers about 2.5 centimeters across.

Habitat and Distribution: Some species are found in rich forests where they are conspicuous before the leaves develop. Western species are found throughout most of the northern United States and in Canada.

Edible Parts: The tubers are edible but you should boil them before eating.

Juniper
Juniperus species

Description: Junipers, sometimes called cedars, are trees or shrubs with very small, scalelike leaves densely crowded around the branches. Each leaf is less than 1.2 centimeters long. All species have a distinct aroma resembling the well-known cedar. The berrylike cones are usually blue and covered with a whitish wax.

Habitat and Distribution: Look for junipers in open, dry, sunny areas throughout North America and northern Europe. Some species are found in southeastern Europe, across Asia to Japan, and in the mountains of North Africa.

Edible Parts: The berries and twigs are edible. Eat the berries raw or roast the seeds to use as a coffee substitute. Use dried and crushed berries as a seasoning for meat. Gather young twigs to make a tea.

⚠ **CAUTION**

Many plants may be called cedars but are not related to junipers and may be harmful. Always look for the berrylike structures, neddle leaves, and resinous, fragrant sap to be sure the plant you have is a juniper.

Lotus
Nelumbo species

Description: There are two species of lotus: one has yellow flowers and the other pink flowers. The flowers are large and showy. The leaves, which may float on or rise above the surface of the water, often reach 1.5 meters in radius. The fruit has a distinctive flattened shape and contains up to 20 hard seeds.

Habitat and Distribution: The yellow-flowered lotus is native to North America. The pink-flowered species, which is widespread in the Orient, is planted in many other areas of the world. Lotuses are found in quiet fresh water.

Edible Parts: All parts of the plant are edible raw or cooked. The underwater parts contain large quantities of starch. Dig the fleshy portions from the mud and bake or boil them. Boil the young leaves and eat them as a vegetable. The seeds have a pleasant flavor and are nutritious. Eat them raw, or parch and grind them into flour.

Malanga
Xanthosoma caracu

Description: This plant has soft, arrow-shaped leaves, up to 60 centimeters long. The leaves have no aboveground stems.

Habitat and Distribution: This plant grows widely in the Caribbean region. Look for it in open, sunny fields.

Edible Parts: The tubers are rich in starch. Cook them before eating to destroy a poison contained in all parts of the plant.

Mango
Mangifera indica

Description: This tree may reach 30 meters in height. It has alternate, simple, shiny, dark green leaves. Its flowers are small and

inconspicuous. Its fruits have a large single seed. There are many cultivated varieties of mango. Some have red flesh, others yellow or orange, often with many fibers and a kerosene taste.

Habitat and Distribution: This tree grows in warm, moist regions. It is native to northern India, Burma, and western Malaysia. It is now grown throughout the tropics.

Edible Parts: The fruits are a nutritious food source. The unripe fruit can be peeled and its flesh eaten by shredding it and eating it like a salad. The ripe fruit can be peeled and eaten raw. Roasted seed kernels are edible.

> ⚠ CAUTION
> If you are sensitive to poison ivy, avoid eating mangoes, as they cause a severe reaction in sensitive individuals.

Manioc
Manihot utillissima

Description: Manioc is a perennial shrubby plant, 1 to 3 meters tall, with jointed stems and deep green, fingerlike leaves. It has large, fleshy rootstocks.

Habitat and Distribution: Manioc is widespread in all tropical climates, particularly in moist areas. Although cultivated extensively, it may be found in abandoned gardens and growing wild in many areas.

Edible Parts: The rootstocks are full of starch and high in food value. Two kinds of manioc are known: bitter and sweet. Both are edible. The bitter type contains poisonous hydrocyanic acid. To prepare manioc, first grind the fresh manioc root into a pulp, then cook it for at least 1 hour to remove the bitter poison from the roots. Then flatten the pulp into cakes and bake as bread. Manioc cakes or flour will keep almost indefinitely if protected against insects and dampness. Wrap them in banana leaves for protection.

⚠ CAUTION
For safety, always cook the roots of either type.

Marsh marigold

Caltha palustris

Description: This plant has rounded, dark green leaves arising from a short stem. It has bright yellow flowers.

Habitat and Distribution: This plant is found in bogs, lakes, and slow-moving streams. It is abundant in arctic and subarctic regions and in much of the eastern region of the northern United States.

Edible Parts: All parts are edible if boiled.

⚠ CAUTION

As with all water plants, do not eat this plant raw. Raw water plants may carry dangerous organisms that are removed only by cooking.

Mulberry

Morus species

Description: This tree has alternate, simple, often lobed leaves with rough surfaces. Its fruits are blue or black and many-seeded.

Habitat and Distribution: Mulberry trees are found in forests, along roadsides, and in abandoned fields in Temperate and Tropical Zones of North America, South America, Europe, Asia, and Africa.

Edible Parts: The fruit is edible raw or cooked. It can be dried for eating later.

> ## ⚠ CAUTION
> When eaten in quantity, mulberry fruit acts as a laxative. Green, unripe fruit can be hallucinogenic and cause extreme nausea and cramps.

Other Uses: You can shred the inner bark of the tree and use it to make twine or cord.

Nettle

Urtica and *Laportea* species

Description: These plants grow several feet high. They have small, inconspicuous flowers. Fine, hairlike bristles cover the stems, leafstalks, and undersides of leaves. The bristles cause a stinging sensation when they touch the skin.

Habitat and Distribution: Nettles prefer moist areas along streams or at the margins of forests. They are found throughout North America, Central America, the Caribbean, and northern Europe.

Edible Parts: Young shoots and leaves are edible. Boiling the plant for 10 to 15 minutes destroys the stinging element of the bristles. This plant is very nutritious.

Other Uses: Mature stems have a fibrous layer that you can divide into individual fibers and use to weave string or twine.

Nips palm
Nips fruticans

Description: This palm has a short, mainly underground trunk and very large, erect leaves up to 6 meters tall. The leaves are divided into leaflets. A flowering head forms on a short erect stern that rises among the palm leaves. The fruiting (seed) head is dark brown and may be 30 centimeters in diameter.

Habitat and Distribution: This palm is common on muddy shores in coastal regions throughout eastern Asia.

Edible Parts: The young flower stalk and the seeds provide a good source of water and food. Cut the flower stalk and collect the juice. The juice is rich in sugar. The seeds are hard but edible.

Other Uses: The leaves are excellent as thatch and coarse weaving material.

Oak
Quercus **species**

Description: Oak trees have alternate leaves and acorn fruits. There are two main groups of oaks: red and white. The red oak group has leaves with bristles and smooth bark in the upper part of the tree. Red oak acorns take 2 years to mature. The white oak group has leaves without bristles and a rough bark in the upper portion of the tree. White oak acorns mature in 1 year.

Habitat and Distribution: Oak trees are found in many habitats throughout North America, Central America, and parts of Europe and Asia.

Edible Parts: All parts are edible, but often contain large quantities of bitter substances. White oak acorns usually have a better flavor than red oak acorns. Gather and shell the acorns. Soak red oak acorns in water for 1 to 2 days to remove the bitter substance. You can speed up this process by putting wood ashes in the water in which you soak the acorns. Boil the acorns or grind them into flour and use the flour for baking. You can use acorns that you baked until very dark as a coffee substitute.

> ⚠ **CAUTION**
> Tannic acid gives the acorns their bitter taste. Eating an excessive amount of acorns high in tannic acid can lead to kidney failure. Before eating acorns, leach out this chemical.

Other Uses: Oak wood is excellent for building or burning. Small oaks can be split and cut into long thin strips (3 to 6 millimeters thick and 1.2 centimeters wide) used to weave mats, baskets, or frameworks for packs, sleds, furniture, etc. Oak bark soaked in water produces a tanning solution used to preserve leather.

Orach

Atriplex **species**

Description: This plant is vinelike in growth and has arrowhead-shaped, alternate leaves up to 5 centimeters long. Young leaves maybe silver-colored. Its flowers and fruits are small and inconspicuous.

Habitat and Distribution: Orach species are entirely restricted to salty soils. They are found along North America's coasts and on the shores of alkaline lakes inland. They are also found along seashores from the Mediterranean countries to inland areas in North Africa and eastward to Turkey and central Siberia.

Edible Parts: The entire plant is edible raw or boiled.

Palmetto palm
Sabal palmetto

Description: The palmetto palm is a tall, unbranched tree with persistent leaf bases on most of the trunk. The leaves are large, simple, and palmately lobed. Its fruits are dark blue or black with a hard seed.

Habitat and Distribution: The palmetto palm is found throughout the coastal regions of the southeastern United States.

Edible Parts: The fruits are edible raw. The hard seeds may be ground into flour. The heart of the palm is a nutritious food source at any time. Cut off the top of the tree to obtain the palm heart.

Papaya or pawpaw
Carica papaya

Description: The papaya is a small tree 1.8 to 6 meters tall, with a soft, hollow trunk. When cut, the entire plant exudes a milky juice. The

trunk is rough and the leaves are crowded at the trunk's apex. The fruit grows directly from the trunk, among and below the leaves. The fruit is green before ripening. When ripe, it turns yellow or remains greenish with a squashlike appearance.

Habitat and Distribution: Papaya is found in rain forests and semi-evergreen seasonal forests in tropical regions and in some temperate regions as well. Look for it in moist areas near clearings and former habitations. It is also found in open, sunny places in uninhabited jungle areas.

Edible Parts: The ripe fruit is high in vitamin C. Eat it raw or cook it like squash. Place green fruit in the sun to make it ripen quickly. Cook the young papaya leaves, flowers, and stems carefully, changing the water as for taro.

⚠ CAUTION
Be careful not to get the milky sap from the unripe fruit into your eyes. It will cause intense pain and temporary—sometimes even permanent—blindness.

Other Uses: Use the milky juice of the unripe fruit to tenderize tough meat. Rub the juice on the meat.

Persimmon
Diospyros virginiana and other species

Description: These trees have alternate, dark green, elliptic leaves with entire margins. The flowers are inconspicuous. The fruits are orange, have a sticky consistency, and have several seeds.

Habitat and Distribution: The persimmon is a common forest margin tree. It is wide spread in Africa, eastern North America, and the Far East.

Edible Parts: The leaves are a good source of vitamin C. The fruits are edible raw or baked. To make tea, dry the leaves and soak them in hot water. You can eat the roasted seeds.

> ⚠ CAUTION
>
> Some persons are unable to digest persimmon pulp. Unripe persimmons are highly astringent and inedible.

Pincushion cactus
Mammilaria species

Description: Members of this cactus group are round, short, barrel-shaped, and without leaves. Sharp spines cover the entire plant.

Habitat and Distribution: These cacti are found throughout much of the desert regions of the western United States and parts of Central America.

Edible Parts: They are a good source of water in the desert.

Pine
Pinus species

Description: Pine trees are easily recognized by their needlelike leaves grouped in bundles. Each bundle may contain one to five needles, the number varying among species. The tree's odor and sticky sap provide a simple way to distinguish pines from similar looking trees with needlelike leaves.

Habitat and Distribution: Pines prefer open, sunny areas. They are found throughout North America, Central America, much of the Caribbean region, North Africa, the Middle East, Europe, and some places in Asia.

Edible Parts: The seeds of all species are edible. You can collect the young male cones, which grow only in the spring, as a survival food. Boil or bake the young cones. The bark of young twigs is edible. Peel off the bark of thin twigs. You can chew the juicy inner bark; it is rich in sugar and vitamins. Eat the seeds raw or cooked. Green pine needle tea is high in vitamin C.

Other Uses: Use the resin to waterproof articles. Also use it as glue. Collect the resin from the tree. If there is not enough resin on the tree, cut a notch in the bark so more sap will seep out. Put the resin in a container and heat it. The hot resin is your glue. Use it as is or add a small amount of ash dust to strengthen it. Use it immediately. You can use hardened pine resin as an emergency dental filling.

Plantain, broad and narrow leaf
Plantago species

Description: The broad leaf plantain has leaves over 2.5 centimeters across that grow close to the ground. The flowers are on a spike that

rises from the middle of the cluster of leaves. The narrow leaf plantain has leaves up to 12 centimeters long and 2.5 centimeters wide, covered with hairs. The leaves form a rosette. The flowers are small and inconspicuous.

Habitat and Distribution: Look for these plants in lawns and along roads in the North Temperate Zone. This plant is a common weed throughout much of the world.

Edible Parts: The young tender leaves are edible raw. Older leaves should be cooked. Seeds are edible raw or roasted.

Other Uses: To relieve pain from wounds and sores, wash and soak the entire plant for a short time and apply it to the injured area. To treat diarrhea, drink tea made from 28 grams (1 ounce) of the plant leaves boiled in 0.5 liter of water. The seeds and seed husks act as laxatives.

Pokeweed
Phytolacca americana

Description: This plant may grow as high as 3 meters. Its leaves are elliptic and up to 1 meter in length. It produces many large clusters of purple fruits in late spring.

Habitat and Distribution: Look for this plant in open, sunny areas in forest clearings, in fields, and along roadsides in eastern North America, Central America, and the Caribbean.

Edible Parts: The young leaves and stems are edible cooked. Boil them twice, discarding the water from the first boiling. The fruits are edible if cooked.

> ⚠ CAUTION
>
> All parts of this plant are poisonous if eaten raw. Never eat the underground portions of the plant as these contain the highest concentrations of the poisons. Do not eat any plant over 25 centimeters tall or when red is showing in the plant.

Other Uses: Use the juice of fresh berries as a dye.

Prickly pear cactus
Opuntia species

Description: This cactus has flat, padlike stems that are green. Many round, furry dots that contain sharp-pointed hairs cover these stems.

Habitat and Distribution: This cactus is found in arid and semiarid regions and in dry, sandy areas of wetter regions throughout most of the United States and Central and South America. Some species are planted in arid and semiarid regions of other parts of the world.

Edible Parts: All parts of the plant are edible. Peel the fruits and eat them fresh or crush them to prepare a refreshing drink. Avoid the tiny, pointed hairs. Roast the seeds and grind them to a flour.

⚠ **CAUTION**
Avoid any prickly pear cactus-like plant with milky sap.

Other Uses: The pad is a good source of water. Peel it carefully to remove all sharp hairs before putting it in your mouth. You can also use the pads to promote healing. Split them and apply the pulp to wounds.

Purslane
Portulaca oleracea

Description: This plant grows close to the ground. It is seldom more than a few centimeters tall. Its stems and leaves are fleshy and often tinged with red. It has paddle-shaped leaves, 2.5 centimeters or less long, clustered at the tips of the stems. Its flowers are yellow or pink. Its seeds are tiny and black.

Habitat and Distribution: It grows in full sun in cultivated fields, field margins, and other weedy areas throughout the world.

Edible Parts: All parts are edible. Wash and boil the plants for a tasty vegetable or eat them raw. Use the seeds as a flour substitute or eat them raw.

Rattan palm
Calamus species

Description: The rattan palm is a stout, robust climber. It has hooks on the midrib of its leaves that it uses to remain attached to trees on which it grows. Sometimes, mature stems grow to 90 meters. It has alternate, compound leaves and a whitish flower.

Habitat and Distribution: The rattan palm is found from tropical Africa through Asia to the East Indies and Australia. It grows mainly in rain forests.

Edible Parts: Rattan palms hold a considerable amount of starch in their young stem tips. You can eat them roasted or raw. In other kinds, a gelatinous pulp, either sweet or sour, surrounds the seeds. You can suck out this pulp. The palm heart is also edible raw or cooked.

Other Uses: You can obtain large amounts of potable water by cutting the ends of the long stems (see Chapter 6). The stems can be used to make baskets and fish traps.

Reed
Phragmites australis

Description: This tall, coarse grass grows to 3.5 meters tall and has gray-green leaves about 4 centimeters wide. It has large masses of brown flower branches in early summer. These rarely produce grain and become fluffy, gray masses late in the season.

Habitat and Distribution: Look for reed in any open, wet area, especially one that has been disturbed through dredging. Reed is found throughout the temperate regions of both the Northern and Southern Hemispheres.

Edible Parts: All parts of the plant are edible raw or cooked in any season. Harvest the stems as they emerge from the soil and boil them. You can also harvest them just before they produce flowers, then dry and beat them into flour. You can also dig up and boil the underground stems, but they are often tough. Seeds are edible raw or boiled, but they are rarely found.

Reindeer moss
Cladonia rangiferina

Description: Reindeer moss is a low-growing plant only a few centimeters tall. It does not flower but does produce bright red reproductive structures.

Habitat and Distribution: Look for this lichen in open, dry areas. It is very common in much of North America.

Edible Parts: The entire plant is edible but has a crunchy, brittle texture. Soak the plant in water with some wood ashes to remove the bitterness, then dry, crush, and add it to milk or to other food.

Rock tripe
Umbilicaria **species**

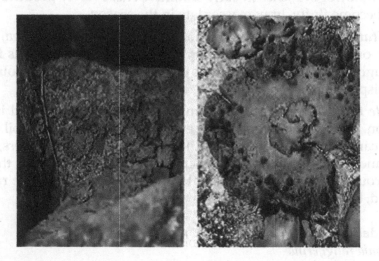

Description: This plant forms large patches with curling edges. The top of the plant is usually black. The underside is lighter in color.

Habitat and Distribution: Look on rocks and boulders for this plant. It is common throughout North America.

Edible Parts: The entire plant is edible. Scrape it off the rock and wash it to remove grit. The plant may be dry and crunchy; soak it in water until it becomes soft. Rock tripes may contain large quantities of bitter substances; soaking or boiling them in several changes of water will remove the bitterness.

⚠ CAUTION
There are some reports of poisoning from rock tripe, so apply the Universal Edibility Test.

Rose apple
Eugenia jambos

Description: This tree grows 3 to 9 meters high. It has opposite, simple, dark green, shiny leaves. When fresh, it has fluffy, yellow-ish-green flowers and red to purple egg-shaped fruit.

Habitat and Distribution: This tree is widely planted in all of the tropics. It can also be found in a semiwild state in thickets, waste places, and secondary forests.

Edible Parts: The entire fruit is edible raw or cooked.

Sago palm
Metroxylon sagu

Description: These palms are low trees, rarely over 9 meters tall, with a stout, spiny trunk. The outer rind is about 5 centimeters thick and hard as bamboo. The rind encloses a spongy inner pith containing a high proportion of starch. It has typical palmlike leaves clustered at the tip.

Habitat and Distribution: Sago palm is found in tropical rain forests. It flourishes in damp lowlands in the Malay Peninsula, New Guinea, Indonesia, the Philippines, and adjacent islands. It is found mainly in swamps and along streams, lakes, and rivers.

Edible Parts: These palms, when available, are of great use to the survivor. One trunk, cut just before it flowers, will yield enough sago to feed a person for 1 year. Obtain sago starch from nonflowering palms. To extract the edible sage, cut away the bark lengthwise from one half of the trunk, and pound the soft, whitish inner part (pith) as fine as possible. Knead the pith in water and strain it through a coarse cloth into a container. The fine, white sago will settle in the container. Once the sago settles, it is ready for use. Squeeze off the excess water and let it dry. Cook it as pancakes or oatmeal. Two kilograms of sago is the nutritional equivalent of 1.5 kilograms of rice. The upper part of the trunk's core does not yield sage, but you can roast it in lumps over a fire. You can also eat the young sago nuts and the growing shoots or palm cabbage.

Other Uses: Use the stems of tall sorghums as thatching materials.

Sassafras
Sassafras albidum

Description: This shrub or small tree bears different leaves on the same plant. Some leaves will have one lobe, some two lobes, and some no lobes. The flowers, which appear in early spring, are small and yellow. The fruits are dark blue. The plant parts have a characteristics root beer smell.

Habitat and Distribution: Sassafras grows at the margins of roads and forests, usually in open, sunny areas. It is a common tree throughout eastern North America.

Edible Parts: The young twigs and leaves are edible fresh or dried. You can add dried young twigs and leaves to soups. Dig the underground portion, peel off the bark, and let it dry. Then boil it in water to prepare sassafras tea.

Other Uses: Shred the tender twigs for use as a toothbrush.

Saxaul
Haloxylon ammondendron

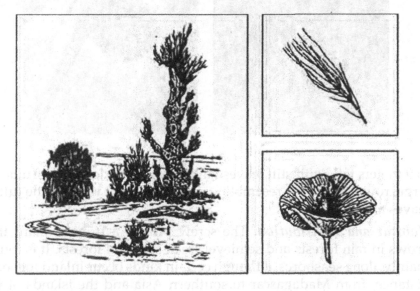

Description: The saxaul is found either as a small tree or as a large shrub with heavy, coarse wood and spongy, water-soaked bark. The branches of the young trees are vivid green and pendulous. The flowers are small and yellow.

Habitat and Distribution: The saxaul is found in desert and arid areas. It is found on the arid salt deserts of Central Asia, particularly in the Turkestan region and east of the Caspian Sea.

Edible Parts: The thick bark acts as a water-storage organ. You can get drinking water by pressing quantities of the bark. This plant is an important source of water in the arid regions in which it grows.

Screw pine
Pandanus species

Description: The screw pine is a strange plant on stilts, or prop roots, that support the plant above-ground so that it appears more or less suspended in midair. These plants are either shrubby or treelike, 3

to 9 meters tall, with stiff leaves having sawlike edges. The fruits are large, roughened balls resembling pineapples, but without the tuft of leaves at the end.

Habitat and Distribution: The screw pine is a tropical plant that grows in rain forests and semievergreen seasonal forests. It is found mainly along seashores, although certain kinds occur inland for some distance, from Madagascar to southern Asia and the islands of the southwestern Pacific. There are about 180 types.

Edible Parts: Knock the ripe fruit to the ground to separate the fruit segments from the hard outer covering. Chew the inner fleshy part. Cook fruit that is not fully ripe in an earth oven. Before cooking, wrap the whole fruit in banana leaves, breadfruit leaves, or any other suitable thick, leathery leaves. After cooking for about 2 hours, you can chew fruit segments like ripe fruit. Green fruit is inedible.

Sea orach
Atriplex halimus

Description: The sea orach is a sparingly branched herbaceous plant with small, gray-colored leaves up to 2.5 centimeters long. Sea orach resembles lamb's quarter, a common weed in most gardens in the United States. It produces its flowers in narrow, densely compacted spikes at the tips of its branches.

Habitat and Distribution: The sea orach is found in highly alkaline and salty areas along seashores from the Mediterranean countries

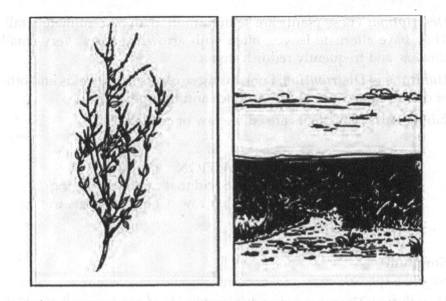

to inland areas in North Africa and eastward to Turkey and central Siberia. Generally, it can be found in tropical scrub and thorn forests, steppes in temperate regions, and most desert scrub and waste areas.

Edible Parts: Its leaves are edible. In the areas where it grows, it has the healthy reputation of being one of the few native plants that can sustain man in times of want.

Sheep sorrel
Rumex acerosella

Description: These plants are seldom more than 30 centimeters tall. They have alternate leaves, often with arrowlike bases, very small flowers, and frequently reddish stems.

Habitat and Distribution: Look for these plants in old fields and other disturbed areas in North America and Europe.

Edible Parts: The plants are edible raw or cooked.

> ### ⚠ CAUTION
> These plants contain oxalic acid that can be damaging if too many plants are eaten raw. Cooking seems to destroy the chemical.

Sorghum
Sorghum species

Description: There are many different kinds of sorghum, all of which bear grains in heads at the top of the plants. The grains are brown, white, red, or black. Sorghum is the main food crop in many parts of the world.

Habitat and Distribution: Sorghum is found worldwide, usually in warmer climates. All species are found in open, sunny areas.

Edible Parts: The grains are edible at any stage of development. When young, the grains are milky and edible raw. Boil the older grains. Sorghum is a nutritious food.

Other Uses: Use the stems of tall sorghum as building materials.

Spatterdock or yellow water lily
Nuphar species

Description: This plant has leaves up to 60 centimeters long with a triangular notch at the base. The shape of the leaves is somewhat variable. The plant's yellow flowers are 2.5 centimeter across and develop into bottle-shaped fruits. The fruits are green when ripe.

Habitat and Distribution: These plants grow throughout most of North America. They are found in quiet, fresh, shallow water (never deeper than 1.8 meters).

Edible Parts: All parts of the plant are edible. The fruits contain several dark brown seeds you can parch or roast and then grind into flour. The large rootstock contains starch. Dig it out of the mud, peel off the outside, and boil the flesh. Sometimes the rootstock contains large quantities of a very bitter compound. Boiling in several changes of water may remove the bitterness.

Sterculia
Sterculia foetida

Description: Sterculias are tall trees, rising in some instances to 30 meters. Their leaves are either undivided or palmately lobed. Their flowers are red or purple. The fruit of all sterculias is similar in aspect, with a red, segmented seedpod containing many edible black seeds.

Habitat and Distribution: There are over 100 species of sterculias distributed through all warm or tropical climates. They are mainly forest trees.

Edible Parts: The large, red pods produce a number of edible seeds. The seeds of all sterculias are edible and have a pleasant taste similar to cocoa. You can eat them like nuts, either raw or roasted.

> ⚠ CAUTION
> Avoid eating large quantities. The seeds may have a laxative effect.

Strawberry
Fragaria species

Description: Strawberry is a small plant with a three-leaved growth pattern. It has small, white flowers usually produced during the spring. Its fruit is red and fleshy.

Habitat and Distribution: Strawberries are found in the North Temperate Zone and also in the high mountains of the southern Western Hemisphere. Strawberries prefer open, sunny areas. They are commonly planted.

Edible Parts: The fruit is edible fresh, cooked, or dried. Strawberries are a good source of vitamin C. You can also eat the plant's leaves or dry them and make a tea with them.

Sugarcane
Saccharum officinarum

Description: This plant grows up to 4.5 meters tall. It is a grass and has grasslike leaves. Its green or reddish stems are swollen where the leaves grow. Cultivated sugarcane seldom flowers.

Habitat and Distribution: Look for sugarcane in fields. It grows only in the tropics (throughout the world). Because it is a crop, it is often found in large numbers.

Edible Parts: The stem is an excellent source of sugar and is very nutritious. Peel the outer portion off with your teeth and eat the sugarcane raw. You can also squeeze juice out of the sugarcane.

Sugar palm
Arenga pinnata

Description: This tree grows about 15 meters high and has huge leaves up to 6 meters long. Needlelike structures stick out of the bases of the leaves. Flowers grow below the leaves and form large conspicuous dusters from which the fruits grow.

Habitat and Distribution: This palm is native to the East Indies but has been planted in many parts off the tropics. It can be found at the margins of forests.

Edible Parts: The chief use of this palm is for sugar. However, its seeds and the tip of its stems are a survival food. Bruise a young flower stalk with a stone or similar object and collect the juice as it comes out. It is an excellent source of sugar. Boil the seeds. Use the tip of the stems as a vegetable.

⚠ CAUTION

The flesh covering the seeds may cause dermatitis.

Other Uses: The shaggy material at the base of the leaves makes an excellent rope as it is strong and resists decay.

Sweetsop
Annona squamosa

Description: This tree is small, seldom more than 6 meters tall, and multi-branched. It has alternate, simple, elongate, dark green leaves. Its fruit is green when ripe, round in shape, and covered with protruding bumps on its surface. The fruit's flesh is white and creamy.

Habitat and Distribution: Look for sweetsop at margins of fields, near villages, and around homesites in tropical regions.

Edible Parts: The fruit flesh is edible raw.

Other Uses: You can use the finely ground seeds as an insecticide.

⚠ CAUTION

The ground seeds are extremely dangerous to the eyes.

Tamarind
Tamarindus indica

Description: The tamarind is a large, densely branched tree, up to 25 meters tall. Its has pinnate leaves (divided like a feather) with 10 to 15 pairs of leaflets.

Habitat and Distribution: The tamarind grows in the drier parts of Africa, Asia, and the Philippines. Although it is thought to be a native of Africa, it has been cultivated in India for so long that it looks like a native tree. It it also found in the American tropics, the West Indies, Central America, and tropical South America.

Edible Parts: The pulp surrounding the seeds is rich in vitamin C and is an important survival food. You can make a pleasantly acid drink by mixing the pulp with water and sugar or honey and letting the mixture mature for several days. Suck the pulp to relieve thirst. Cook the young, unripe fruits or seedpods with meat. Use the young leaves in soup. You must cook the seeds. Roast them above a fire or in ashes. Another way is to remove the seed coat and soak the seeds in salted water and grated coconut for 24 hours, then cook them. You can peel the tamarind bark and chew it.

Taro, cocoyam, elephant ears, eddo, dasheen
Colocasia and Alocasia species

Description: All plants in these groups have large leaves, sometimes up to 1.8 meters tall, that grow from a very short stem. The rootstock is thick and fleshy and filled with starch.

Habitat and Distribution: These plants grow in the humid tropics. Look for them in fields and near homesites and villages.

Edible Parts: All parts of the plant are edible when boiled or roasted. When boiling, change the water once to get rid of any poison.

⚠ CAUTION

If eaten raw, these plants will cause a serious inflammation of the mouth and throat.

Thistle
Cirsium species

Description: This plant may grow as high as 1.5 meters. Its leaves are long-pointed, deeply lobed, and prickly.

Habitat and Distribution: Thistles grow worldwide in dry woods and fields.

Edible Parts: Peel the stalks, cut them into short sections, and boil them before eating. The roots are edible raw or cooked.

> ⚠ CAUTION
> Some thistle species are poisonous.

Other Uses: Twist the tough fibers of the stems to make a strong twine.

Ti
Cordyline terminalis

Description: The ti has unbranched stems with straplike leaves often clustered at the tip of the stem. The leaves vary in color and may be green or reddish. The flowers grow at the plant's top in large, plume-like clusters. The ti may grow up to 4.5 meters tall.

Habitat and Distribution: Look for this plant at the margins of forests or near home-sites in tropical areas. It is native to the Far East but is now widely planted in tropical areas worldwide.

Edible Parts: The roots and very tender young leaves are good survival food. Boil or bake the short, stout roots found at the base of the plant. They are a valuable source of starch. Boil the very young leaves to eat. You can use the leaves to wrap other food to cook over coals or to steam.

Other Uses: Use the leaves to cover shelters or to make a rain cloak. Cut the leaves into liners for shoes; this works especially well if you have a blister. Fashion temporary sandals from the ti leaves. The terminal leaf, if not completely unfurled, can be used as a sterile bandage. Cut the leaves into strips, then braid the strips into rope.

Tree fern
Various genera

Description: Tree ferns are tall trees with long, slender trunks that often have a very rough, barklike covering. Large, lacy leaves uncoil from the top of the trunk.

Habitat and Distribution: Tree ferns are found in wet, tropical forests.

Edible Parts: The young leaves and the soft inner portion of the trunk are edible. Boil the young leaves and eat as greens. Eat the inner portion of the trunk raw or bake it.

Tropical almond
Terminalia catappa

Description: This tree grows up to 9 meters tall. Its leaves are evergreen, leathery, 45 centimeters long, 15 centimeters wide, and very shiny. It has small, yellowish-green flowers. Its fruit is flat, 10 centimeters long, and not quite as wide. The fruit is green when ripe.

Habitat and Distribution: This tree is usually found growing near the ocean. It is a common and often abundant tree in the Caribbean and Central and South America. It is also found in the tropical rain forests of southeastern Asia, northern Australia, and Polynesia.

Edible Parts: The seed is a good source of food. Remove the fleshy, green covering and eat the seed raw or cooked.

Walnut
Juglans species

Description: Walnuts grow on very large trees, often reaching 18 meters tall. The divided leaves characterize all walnut spades. The walnut itself has a thick outer husk that must be removed to reach the hard inner shell of the nut.

Habitat and Distribution: The English walnut, in the wild state, is found from southeastern Europe across Asia to China and is abundant in the Himalayas. Several other species of walnut are found in China and Japan. The black walnut is common in the eastern United States.

Edible Parts: The nut kernel ripens in the autumn. You get the walnut meat by cracking the shell. Walnut meats are highly nutritious because of their protein and oil content.

Other Uses: You can boil walnuts and use the juice as an antifungal agent. The husks of "green" walnuts produce a dark brown dye for clothing or camouflage. Crush the husks of "green" black walnuts and sprinkle them into sluggish water or ponds for use as fish poison.

Water chestnut
Trapa natans

Description: The water chestnut is an aquatic plant that roots in the mud and has finely divided leaves that grow underwater. Its floating leaves are much larger and coarsely toothed. The fruits, borne underwater, have four sharp spines on them. Habitat and

Distribution: The water chestnut is a freshwater plant only. It is a native of Asia but has spread to many parts of the world in both temperate and tropical areas.

Edible Parts: The fruits are edible raw and cooked. The seeds are also a source of food.

Water lettuce
Ceratopteris species

Description: The leaves of water lettuce are much like lettuce and are very tender and succulent. One of the easiest ways of distinguishing water lettuce is by the little plantlets that grow from the margins of the leaves. These little plantlets grow in the shape of a rosette. Water lettuce plants often cover large areas in the regions where they are found.

Habitat and Distribution: Found in the tropics throughout the Old World in both Africa and Asia. Another kind is found in the New World tropics from Florida to South America. Water lettuce grows only in very wet places and often as a floating water plant. Look for water lettuce in still lakes, ponds, and the backwaters of rivers.

Edible Parts: Eat the fresh leaves like lettuce. Be careful not to dip the leaves in the contaminated water in which they are growing. Eat only the leaves that are well out of the water.

⚠ CAUTION
This plant has carcinogenic properties and should only be used as a last resort.

Water lily
Nymphaea odorata

Description: These plants have large, triangular leaves that float on the water's surface, large, fragrant flowers that are usually white, or red, and thick, fleshy rhizomes that grow in the mud.

Habitat and Distribution: Water lilies are found throughout much of the temperate and subtropical regions.

Edible Parts: The flowers, seeds, and rhizomes are edible raw or cooked. To prepare rhizomes for eating, peel off the corky rind. Eat raw, or slice thinly, allow to dry, and then grind into flour. Dry, parch, and grind the seeds into flour.

Other Uses: Use the liquid resulting from boiling the thickened root in water as a medicine for diarrhea and as a gargle for sore throats.

Water plantain
Alisma plantago-aquatica

Description: This plant has small, white flowers and heart-shaped leaves with pointed tips. The leaves are clustered at the base of the plant.

Habitat and Distribution: Look for this plant in fresh water and in wet, full sun areas in Temperate and Tropical Zones.

Edible Parts: The rootstocks are a good source of starch. Boil or soak them in water to remove the bitter taste.

⚠ CAUTION
To avoid parasites, always cook aquatic plants.

Wild caper
Capparis aphylla

Description: This is a thorny shrub that loses its leaves during the dry season. Its stems are gray-green and its flowers pink.

Habitat and Distribution: These shrubs form large stands in scrub and thorn forests and in desert scrub and waste. They are common throughout North Africa and the Middle East.

Edible Parts: The fruit and the buds of young shoots are edible raw.

Wild crab apple or wild apple
Malus species

Description: Most wild apples look enough like domestic apples that the survivor can easily recognize them. Wild apple varieties are much smaller than cultivated kinds; the largest kinds usually do not exceed 5 to 7.5 centimeters in diameter, and most often less. They have small, alternate, simple leaves and often have thorns. Their flowers are white or pink and their fruits reddish or yellowish.

Habitat and Distribution: They are found in the savanna regions of the tropics. In temperate areas, wild apple varieties are found mainly in forested areas. Most frequently, they are found on the edge of woods or in fields. They are found throughout the Northern Hemisphere.

Edible Parts: Prepare wild apples for eating in the same manner as cultivated kinds. Eat them fresh, when ripe, or cooked. Should you need to store food, cut the apples into thin slices and dry them. They are a good source of vitamins.

⚠ CAUTION
Apple seeds contain cyanide compounds. Do not eat.

Wild desert gourd or colocynth
Citrullus colocynthis

Description: The wild desert gourd, a member of the watermelon family, produces an 2.4- to 3-meter-long ground-trailing vine. The perfectly round gourds are as large as an orange. They are yellow when ripe.

Habitat and Distribution: This creeping plant can be found in any climatic zone, generally in desert scrub and waste areas. It grows abundantly in the Sahara, in many Arab countries, on the southeastern coast of India, and on some of the islands of the Aegean Sea. The wild desert gourd will grow in the hottest localities.

Edible Parts: The seeds inside the ripe gourd are edible after they are completely separated from the very bitter pulp. Roast or boil the seeds – their kernels are rich in oil. The flowers are edible. The succulent stem tips can be chewed to obtain water.

Wild dock and wild sorrel
Rumex crispus and Rumex acetosella

Description: Wild dock is a stout plant with most of its leaves at the base of its stem that is commonly 15 to 30 centimeters brig. The plants usually develop from a strong, fleshy, carrotlike taproot. Its flowers are usually very small, growing in green to purplish plumelike clusters. Wild sorrel is similar to the wild dock but smaller. Many of the basal leaves are arrow-shaped but smaller than those of the dock and contain a sour juice.

Habitat and Distribution: These plants can be found in almost all climatic zones of the world, in areas of high as well as low rainfall. Many kinds are found as weeds in fields, along roadsides, and in waste places.

Edible Parts: Because of the tender nature of the foliage, the sorrel and the dock are useful plants, especially in desert areas. You can eat their succulent leaves fresh or slightly cooked. To take away the strong taste, change the water once or twice during cooking. This latter tip is a useful hint in preparing many kinds of wild greens.

Wild fig
Ficus species

Description: These trees have alternate, simple leaves with entire margins. Often, the leaves are dark green and shiny. All figs have a milky, sticky juice. The fruits vary in size depending on the species, but are usually yellow-brown when ripe.

Habitat and Distribution: Figs are plants of the tropics and semi-tropics. They grow in several different habitats, including dense forests, margins of forests, and around human settlements.

Edible Parts: The fruits are edible raw or cooked. Some figs have little flavor.

Wild gourd or luffa sponge
Luffa cylindrica

Description: The luffa sponge is widely distributed and fairly typical of a wild squash. There are several dozen kinds of wild squashes in tropical regions. Like most squashes, the luffa is a vine with leaves 7.5 to 20 centimeters across having 3 lobes. Some squashes have leaves twice this size. Luffa fruits are oblong or cylindrical, smooth, and many-seeded. Luffa flowers are bright yellow. The luffa fruit, when mature, is brown and resembles the cucumber.

Habitat and Distribution: A member of the squash family, which also includes the watermelon, cantaloupe, and cucumber, the luffa sponge is widely cultivated throughout the Tropical Zone. It may be found in a semiwild state in old clearings and abandoned gardens in rain forests and semi-evergreen seasonal forests.

Edible Parts: You can boil the young green (half-ripe) fruit and eat them as a vegetable. Adding coconut milk will improve the flavor. After ripening, the luffa sponge develops an inedible spongelike texture in the interior of the fruit. You can also eat the tender shoots, flowers, and young leaves after cooking them. Roast the mature seeds a little and eat them like peanuts.

Wild grape vine
Vitis species

Description: The wild grape vine climbs with the aid of tendrils. Most grape vines produce deeply lobed leaves similar to the cultivated grape. Wild grapes grow in pyramidal, hanging bunches and are black-blue to amber, or white when ripe.

Habitat and Distribution: Wild grapes are distributed worldwide. Some kinds are found in deserts, others in temperate forests, and others in tropical areas. Wild grapes are commonly found throughout the eastern United States as well as in the southwestern desert areas. Most kinds are rampant climbers over other vegetation. The best place to look for wild grapes is on the edges of forested areas. Wild grapes are also found in Mexico. In the Old World, wild grapes are found from the Mediterranean region eastward through Asia, the East Indies, and to Australia. Africa also has several kinds of wild grapes.

Edible Parts: The ripe grape is the portion eaten. Grapes are rich in natural sugars and, for this reason, are much sought after as a source of energy-giving wild food. None are poisonous.

Other Uses: You can obtain water from severed grape vine stems. Cut off the vine at the bottom and place the cut end in a container. Make a slant-wise cut into the vine about 1.8 meters up on the hanging part. This cut will allow water to flow from the bottom end. As water diminishes in volume, make additional cuts further down the vine.

⚠ CAUTION

To avoid poisoning, do not eat grapelike fruits with only a single seed (moonseed).

Wild onion and garlic
Allium species

Description: Allium cernuum is an example of the many species of wild onions and garlics, all easily recognized by their distinctive odor.

Habitat and Distribution: Wild onions and garlics are found in open, sunny areas throughout the temperate regions. Cultivated varieties are found anywhere in the world.

Edible Parts: The bulbs and young leaves are edible raw or cooked. Use in soup or to flavor meat.

⚠ CAUTION

There are several plants with onionlike bulbs that are extremely poisonous. Be certain that the plant you are using is a true onion or garlic. Do not eat bulbs with no onion smell.

Wild pistachio
Pistacia species

Description: Some kinds of pistachio trees are evergreen, while others lose their leaves during the dry season. The leaves alternate on the stem and have either three large leaves or a number of leaflets. The fruits or nuts are usually hard and dry at maturity.

Habitat and Distribution: About seven kinds of wild pistachio nuts are found in desert, or semidesert areas surrounding the

Mediterranean Sea to Turkey and Afghanistan. It is generally found in evergreen scrub forests or scrub and thorn forests.

Edible Parts: You can eat the oil nut kernels after parching them over coals.

Wild rice
Zizania aquatica

Description: Wild rice is a tall grass that averages 1 to 1.5 meters in height, but may reach 4.5 meters. Its grain grows in very loose heads at the top of the plant and is dark brown or blackish when ripe.

Habitat and Distribution: Wild rice grows only in very wet areas in tropical and temperate regions.

Edible Parts: During the spring and summer, the central portion of the lower sterns and root shoots are edible. Remove the tough covering before eating. During the late summer and fail, collect the straw-covered husks. Dry and parch the husks, break them, and remove the rice. Boil or roast the rice and then beat it into flour.

Wild rose
Rosa species

Description: This shrub grows 60 centimeters to 2.5 meters high. It has alternate leaves and sharp prickles. Its flowers may be red, pink, or yellow. Its fruit, called rose hip, stays on the shrub year-round.

Habitat and Distribution: Look for wild roses in dry fields and open woods throughout the Northern Hemisphere.

Edible Parts: The flowers and buds are edible raw or boiled. In an emergency, you can peel and eat the young shoots. You can boil fresh, young leaves in water to make a tea. After the flower petals fall, eat the rose hips; the pulp is highly nutritious and an excellent source of vitamin C. Crush or grind dried rose hips to make flour.

> ⚠ CAUTION
> Eat only the outer portion of the fruit as the seeds of some species are quite prickly and can cause internal distress.

Wood sorrel
Oxalis species

Description: Wood sorrel resembles shamrock or four-leaf clover, with a bell-shaped pink, yellow, or white flower.

Habitat and Distribution: Wood sorrel is found in Temperate Zones worldwide, in lawns, open areas, and sunny woods.

Edible Parts: Cook the entire plant.

> ### ⚠ CAUTION
> Eat only small amounts of this plant as it contains a fairly high concentration of oxalic acid that can be harmful.

Yam
Dioscorea species

Description: These plants are vines that creep along the ground. They have alternate, heart- or arrow-shaped leaves. Their rootstock may be very large and weigh many kilograms.

Habitat and Distribution: True yams are restricted to tropical regions where they are an important food crop. Look for yams in fields, clearings, and abandoned gardens. They are found in rain forests,

semi-evergreen seasonal forests, and scrub and thorn forests in the tropics. In warm temperate areas, they are found in seasonal hardwood or mixed hardwood-coniferous forests, as well as some mountainous areas.

Edible Parts: Boil the rootstock and eat it as a vegetable.

Yam bean
Pachyrhizus erosus

Description: The yam bean is a climbing plant of the bean family, with alternate, three-parted leaves and a turniplike root. The bluish or purplish flowers are pealike in shape. The plants are often so rampant that they cover the vegetation upon which they are growing.

Habitat and Distribution: The yam bean is native to the American tropics, but it was carried by man years ago to Asia and the Pacific islands. Now it is commonly cultivated in these places, and is also found growing wild in forested areas. This plant grows in wet areas of tropical regions.

Edible Parts: The tubers are about the size of a turnip and they are crisp, sweet, and juicy and have a nutty flavor. They are nourishing and at the same time quench the thirst. Eat them raw or boiled. To make flour, slice the raw tubers, let them dry in the sun, and grind into a flour that is high in starch and may be used to thicken soup.

⚠ CAUTION

The raw seeds are poisonous.

CHAPTER 7

POISONOUS PLANTS

Successful use of plants in a survival situation depends on positive identification. Knowing poisonous plants is as important to a survivor as knowing edible plants. Knowing the poisonous plants will help you avoid sustaining injuries from them.

HOW PLANTS POISON

Plants generally poison by—

- Ingestion. When a person eats a part of a poisonous plant.
- Contact. When a person makes contact with a poisonous plant that causes any type of skin irritation or dermatitis.
- Absorption or inhalation. When a person either absorbs the poison through the skin or inhales it into the respiratory system.

Plant poisoning ranges from minor irritation to death. A common question asked is, "How poisonous is this plant?" It is difficult to say how poisonous plants are because—

- Some plants require contact with a large amount of the plant before noticing any adverse reaction while others will cause death with only a small amount.
- Every plant will vary in the amount of toxins it contains due to different growing conditions and slight variations in subspecies.
- Every person has a different level of resistance to toxic substances.
- Some persons may be more sensitive to a particular plant.

Some common misconceptions about poisonous plants are—

- *Watch the animals and eat what they eat.* Most of the time this statement is true, but some animals can eat plants that are poisonous to humans.

- *Boil the plant in water and any poisons will be removed.* Boiling removes many poisons, but not all.
- *Plants with a red color are poisonous.* Some plants that are red are poisonous, but not all.

The point is there is no one rule to aid in identifying poisonous plants. You must make an effort to learn as much about them as possible.

ALL ABOUT PLANTS

It is to your benefit to learn as much about plants as possible. Many poisonous plants look like their edible relatives or like other edible plants. For example, poison hemlock appears very similar to wild carrot. Certain plants are safe to eat in certain seasons or stages of growth and poisonous in other stages. For example, the leaves of the pokeweed are edible when it first starts to grow, but it soon becomes poisonous. You can eat some plants and their fruits only when they are ripe. For example, the ripe fruit of mayapple is edible, but all other parts and the green fruit are poisonous. Some plants contain both edible and poisonous parts; potatoes and tomatoes are common plant foods, but their green parts are poisonous.

Some plants become toxic after wilting. For example, when the black cherry starts to wilt, hydrocyanic acid develops. Specific preparation methods make some plants edible that are poisonous raw. You can eat the thinly sliced and thoroughly dried corms (drying may take a year) of the jack-in-the-pulpit, but they are poisonous if not thoroughly dried.

Learn to identify and use plants before a survival situation. Some sources of information about plants are pamphlets, books, films, nature trails, botanical gardens, local markets, and local natives. Gather and cross-reference information from as many sources as possible, because many sources will not contain all the information needed.

RULES FOR AVOIDING POISONOUS PLANTS

Your best policy is to be able to look at a plant and identify it with absolute certainty and to know its uses or dangers. Many times this is not possible. If you have little or no knowledge of the local vegetation, use the rules to select plants for the "Universal Edibility Test." Remember, avoid —

- All mushrooms. Mushroom identification is very difficult and must be precise, even more so than with other plants. Some

mushrooms cause death very quickly. Some mushrooms have no known antidote. Two general types of mushroom poisoning are gastrointestinal and central nervous system.
- Contact with or touching plants unnecessarily.

CONTACT DERMATITIS

Contact dermatitis from plants will usually cause the most trouble in the field. The effects may be persistent, spread by scratching, and are particularly dangerous if there is contact in or around the eyes.

The principal toxin of these plants is usually an oil that gets on the skin upon contact with the plant. The oil can also get on equipment and then infect whoever touches the equipment. Never burn a contact poisonous plant because the smoke may be as harmful as the plant. There is a greater danger of being affected when overheated and sweating. The infection may be local or it may spread over the body.

Symptoms may take from a few hours to several days to appear. Signs and symptoms can include burning, reddening, itching, swelling, and blisters.

When you first contact the poisonous plants or the first symptoms appear, try to remove the oil by washing with soap and cold water. If water is not available, wipe your skin repeatedly with dirt or sand. Do not use dirt if blisters have developed. The dirt may break open the blisters and leave the body open to infection. After you have removed the oil, dry the area. You can wash with a tannic acid solution and crush and rub jewelweed on the affected area to treat plant-caused rashes. You can make tannic acid from oak bark.

Poisonous plants that cause contact dermatitis are—

- Cowhage.
- Poison ivy.
- Poison oak.
- Poison sumac.
- Rengas tree.
- Trumpet vine.

INGESTION POISONING

Ingestion poisoning can be very serious and could lead to death very quickly. Do not eat any plant unless you have positively identified it first. Keep a log of all plants eaten.

Signs and symptoms of ingestion poisoning can include nausea, vomiting, diarrhea, abdominal cramps, depressed heartbeat and respiration, headaches, hallucinations, dry mouth, unconsciousness, coma, and death.

If you suspect plant poisoning, try to remove the poisonous material from the victim's mouth and stomach as soon as possible. Induce vomiting by tickling the back of his throat or by giving him warm saltwater, if he is conscious. Dilute the poison by administering large quantities of water or milk, if he is conscious.

The following plants can cause ingestion poisoning if eaten:

- Castor bean.
- Chinaberry.
- Death camas.
- Lantana.
- Manchineel.
- Oleander.
- Pangi.
- Physic nut.
- Poison and water hemlocks.
- Rosary pea.
- Strychnine tree.

POISONOUS PLANTS

Castor bean, castor-oil plant, palma Christi
Ricinus communis
Spurge (*Euphorbiaceae*) Family

Description: The castor bean is a semiwoody plant with large, alternate, starlike leaves that grows as a tree in tropical regions and as an annual in temperate regions. Its flowers are very small and inconspicuous. Its fruits grow in clusters at the tops of the plants.

⚠ CAUTION
All parts of the plant are very poisonous to eat. The seeds are large and may be mistaken for a beanlike food.

Habitat and Distribution: This plant is found in all tropical regions and has been introduced to temperate regions.

Chinaberry
Melia azedarach
Mahogany (*Meliaceae*) Family

Description: This tree has a spreading crown and grows up to 14 meters tall. It has alternate, compound leaves with toothed leaflets. Its flowers are light purple with a dark center and grow in ball-like masses. It has marble-sized fruits that are light orange when first formed but turn lighter as they become older.

> ⚠ CAUTION
> All parts of the tree should be considered dangerous
> if eaten. Its leaves are a natural insecticide and will
> repel insects from stored fruits and grains. Take care
> not to eat leaves mixed with the stored food.

Habitat and Distribution: Chinaberry is native to the Himalayas and eastern Asia but is now planted as an ornamental tree throughout the tropical and subtropical regions. It has been introduced to the southern United States and has escaped to thickets, old fields, and disturbed areas.

Cowhage, cowage, cowitch
Mucuna pruritum
Leguminosae (*Fabaceae*) Family

Description: A vinelike plant that has oval leaflets in groups of three and hairy spikes with dull purplish flowers. The seeds are brown, hairy pods.

> ⚠ CAUTION
> Contact with the pods and flowers causes irritation
> and blindness if in the eyes.

Death camas, death lily
Zigadenus species
Lily (*Liliaceae*) Family

Description: This plant arises from a bulb and may be mistaken for an onionlike plant. Its leaves are grasslike. Its flowers are six-parted and the petals have a green, heart-shaped structure on them. The flowers grow on showy stalks above the leaves.

⚠ CAUTION
All parts of this plant are very poisonous. Death camas does not have the onion smell.

Habitat and Distribution: Death camas is found in wet, open, sunny habitats, although some species favor dry, rocky slopes. They are common in parts of the western United States. Some species are found in the eastern United States and in parts of the North American western subarctic and eastern Siberia.

Lantana
Lantana camara
Vervain (*Verbenaceae*) Family

Description: Lantana is a shrublike plant that may grow up to 45 centimeters high. It has opposite, round leaves and flowers borne in

flat-topped clusters. The flower color (which varies in different areas) maybe white, yellow, orange, pink, or red. It has a dark blue or black berrylike fruit. A distinctive feature of all parts of this plant is its strong scent.

⚠ CAUTION
All parts of this plant are poisonous if eaten and can be fatal. This plant causes dermatitis in some individuals.

Habitat and Distribution: Lantana is grown as an ornamental in tropical and temperate areas and has escaped cultivation as a weed along roads and old fields.

Manchineel
Hippomane mancinella
Spurge (*Euphorbiaceae*) Family

Description: Manchineel is a tree reaching up to 15 meters high with alternate, shiny green leaves and spikes of small greenish flowers. Its fruits are green or greenish-yellow when ripe.

> ⚠ **CAUTION**
>
> This tree is extremely toxic. It causes severe dermatitis in most individuals after only half an hour. Even water dripping from the leaves may cause dermatitis. The smoke from burning it irritates the eyes. No part of this plant should be considered a food.

Habitat and Distribution: The tree prefers coastal regions. Found in south Florida, the Caribbean, Central America, and northern South America.

Oleander
Nerium oleander
Dogbane (*Apocynaceae*) Family

Description: This shrub or small tree grows to about 9 meters, with alternate, very straight, dark green leaves. Its flowers may be white, yellow, red, pink, or intermediate colors. Its fruit is a brown, podlike structure with many small seeds.

> ⚠ **CAUTION**
>
> All parts of the plant are very poisonous. Do not use the wood for cooking; it gives off poisonous fumes that can poison food.

Habitat and Distribution: This native of the Mediterranean area is now grown as an ornamental in tropical and temperate regions.

Pangi
Pangium edule
Pangi Family

Description: This tree, with heart-shaped leaves in spirals, reaches a height of 18 meters. Its flowers grow in spikes and are green in color. Its large, brownish, pear-shaped fruits grow in clusters.

⚠ CAUTION
All parts are poisonous, especially the fruit.

Habitat and Distribution: Pangi trees grow in southeast Asia

Physic nut
Jatropha curcas
Spurge (*Euphoriaceae*) Family

Description: This shrub or small tree has large, 3- to 5-parted alternate leaves. It has small, greenish-yelllow flowers and its yellow, apple-sized fruits contain three large seeds.

⚠ **CAUTION**
The seeds taste sweet but their oil is violently purgative. All parts of the physic nut are poisonous.

Habitat and Distribution: Throughout the tropics and southern United States.

Poison hemlock, fool's parsley
Conium maculatum
Parsley (*Apiaceae*) Family

Description: This biennial herb may grow to 2.5 meters high. The smooth, hollow stem may or may not be purple or red striped or mottled. Its white flowers are small and grow in small groups that tend to form flat umbels. Its long, turniplike taproot is solid.

> ⚠ **CAUTION**
>
> This plant is very poisonous and even a very small amount may cause death. This plant is easy to confuse with wild carrot or Queen Anne's lace, especially in its first stage of growth. Wild carrot or Queen Anne's lace has hairy leaves and stems and smells like carrot. Poison hemlock does not.

Habitat and Distribution: Poison hemlock grows in wet or moist ground like swamps, wet meadows, stream banks, and ditches. Native to Eurasia, it has been introduced to the United States and Canada.

Poison ivy and poison oak
Toxicodendron radicans and Toxicodendron diversibba
Cashew (*Anacardiacese*) Family

Description: These two plants are quite similar in appearance and will often crossbreed to make a hybrid. Both have alternate, compound leaves with three leaflets. The leaves of poison ivy are smooth or serrated. Poison oak's leaves are lobed and resemble oak leaves.

Poison ivy grows as a vine along the ground or climbs by red feeder roots. Poison oak grows like a bush. The greenish-white flowers are small and inconspicuous and are followed by waxy green berries that turn waxy white or yellow, then gray.

⚠ CAUTION

All parts, at all times of the year, can cause serious contact dermatitis.

Habitat and Distribution: Poison ivy and oak can be found in almost any habitat in North America.

Poison sumac
Toxicodendron vernix
Cashew (*Anacardiacese*) Family

Description: Poison sumac is a shrub that grows to 8.5 meters tall. It has alternate, pinnately compound leafstalks with 7 to 13 leaflets. Flowers are greenish-yellow and inconspicuous and are followed by white or pale yellow berries.

⚠ CAUTION

All parts can cause serious contact dermatitis at all times of the year.

Habitat and Distribution: Poison sumac grows only in wet, acid swamps in North America.

Renghas tree, rengas tree, marking nut, black-varnish tree
Gluta
Cashew (*Anacardiacese*) Family

Renghas tree, rengas tree, marking nut, black-varnish tree
Gluta
Cashew (*Anacardiaceae*) Family

Description: This family comprises about 48 species of trees or shrubs with alternating leaves in terminal or axillary panicles. Flowers are similar to those of poison ivy and oak.

> **CAUTION**
> Can cause contact dermatitis similar to poison ivy or poison oak.

Habitat and Distribution: India, east to Southeast Asia.

Description: This family comprises about 48 species of trees or shrubs with alternating leaves in terminal or axillary panicles. Flowers are similar to those of poison ivy and oak.

⚠ **CAUTION**
Can cause contact dermatitis similar to poison ivy and oak.

Habitat and Distribution: India, east to Southeast Asia.

Rosary pea or crab's eyes
Abrus precatorius
Leguminosae (*Fabaceae*) Family

Description: This plant is a vine with alternate compound leaves, light purple flowers, and beautiful seeds that are red and black.

Habitat and Distribution: This is a common weed in parts of Africa, southern Florida, Hawaii, Guam, the Caribbean, and Central and South America.

⚠ CAUTION

This plant is one of the most dangerous plants. One seed may contain enough poison to kill an adult.

Strychnine tree
Nux vomica
Logania (*Loganiaceae*) Family

Description: The strychnine tree is a medium-sized evergreen, reaching a height of about 12 meters, with a thick, frequently crooked trunk. Its deeply veined oval leaves grow in alternate pairs. Small,

loose clusters of greenish flowers appear at the ends of branches and are followed by fleshy, orange-red berries about 4 centimeters in diameter.

> ⚠ **CAUTION**
> The berries contain the disclike seeds that yield the poisonous substance strychnine. All parts of the plant are poisonous.

Habitat and Distribution: A native of the tropics and subtropics of southeastern Asia and Australia.

Trumpet vine or trumpet creeper
Campsis radicans
Trumpet creeper (*Bignoniaceae*) Family

Description: This woody vine may climb to 15 meters high. It has pealike fruit capsules. The leaves are pinnately compound, 7 to 11 toothed leaves per leaf stock. The trumpet-shaped flowers are orange to scarlet in color.

> ⚠ **CAUTION**
> This plant causes contact dermatitis.

Habitat and Distribution: This vine is found in wet woods and thickets throughout eastern and central North America.

Water hemlock or spotted cowbane
Cicuta maculata
Parsley (*Apiaceae*) Family

Description: This perennial herb may grow to 1.8 meters high. The stem is hollow and sectioned off like bamboo. It may or may not be purple or red striped or mottled. Its flowers are small, white, and grow in groups that tend to form flat umbels. Its roots may have hollow air chambers and, when cut, may produce drops of yellow oil.

> ⚠ **CAUTION**
> This plant is very poisonous and even a very small amount of this plant may cause death. Its roots have been mistaken for parsnips.

Habitat and Distribution: Water hemlock grows in wet or moist ground like swamps, wet meadows, stream banks, and ditches throughout the Unites States and Canada.